Hard and Soft Acids and Bases Principle in Organic Chemistry

Hard and Soft Acids and Bases Principle in Organic Chemistry

TSE-LOK HO

Department of Chemistry
Case Western Reserve University
Cleveland, Ohio

ACADEMIC PRESS   New York   San Francisco   London   1977
A Subsidiary of Harcourt Brace Jovanovich, Publishers

ACADEMIC PRESS, INC.
111 Fifth Avenue, New York, New York 10003

United Kingdom Edition published by
ACADEMIC PRESS, INC. (LONDON) LTD.
24/28 Oval Road. London NW1

Library of Congress Cataloging in Publication Data

Ho, Tse-Lok.
    Hard and soft acids and bases principle in organic
chemistry
    Includes bibliographical references and index.
    1.   Acids.   2.   Bases (Chemistry)   3.   Chemistry
Organic.   4.   Chemical reactions.   I.   Title.
QD477.H6   547'.1'39   76-13939
ISBN 0−12−350050−8

To
Honor, Jocelyn,
and
my Parents

Contents

Preface

The formulation of a new concept always signifies major advances in scientific endeavors, and such was the case when the hard and soft acids and bases (HSAB) principle was introduced by Professor R. G. Pearson more than a dozen years ago. Since that time a tremendous number of chemical phenomena related directly to or derived from the HSAB principle have accumulated, most of which support Pearson's original premise. As it stands, there is no denying the value of the principle as a convenient yet powerful tool for assessing and predicting chemical events without having to resort to lengthy and cumbersome calculation. Because the HSAB principle offers great assistance in and stimulation for both pedagogy and research, it should be introduced to students of chemistry at an early stage.

Initially used to explain inorganic coordination chemistry, HSAB has subsequently been applied to almost every chemical ramification with considerable success. Although pioneering applications to the organic area by Hudson, Saville, and Pearson himself have irrefutably demonstrated the relevance of HSAB to organic chemistry, subject matter hitherto touched on can be likened to the exposed tip of an iceberg. Pertinent findings are bound to surface as long as organic research does not cease.

This volume represents an attempt to examine the many organic facets within the HSAB context. It is descriptive in nature, and is intended for the senior undergraduate or graduate chemistry major. However, it can also serve as an introduction and reference source for organic chemists who are not familiar with the HSAB concept. Although data quoted in this monograph are established fact, a good deal of it is related to the HSAB principle for the first time.

A number of chemists have been bothered by the terms "hard" and "soft." The irritation of these critics may be somewhat soothed now that physicists describe the supposedly elementary particles, quarks, in terms of "color," "flavor," and "charm" $\{$S. L. Glashow, *Scientific American* **233** [4], 38 (1975)$\}$.

I am keenly aware of the fact that a survey of all aspects is an insuperable task, thus personal predilection and ignorance became factors dictating selection or omission of various topics. Where the original researcher may not agree with the treatment presented, I assume total responsibility for any misrepresentation. I hope that some of the material presented will initiate further chemical research along these lines.

Tse-Lok Ho*

* Present address: Medical Research Department, Brookhaven National Laboratory, Upton, New York.

1

Introduction

1.1. DEFINITIONS OF ACIDS AND BASES

Classically acids are defined as substances that give off protons and bases are those that generate hydroxide ions when dissolved in water. These restrictive definitions are no longer adequate and have been modified or supplemented by others. Currently the most widely used acid–base definitions are due to Brønsted and Lowry, and Lewis.

1.1.1. The Brønsted–Lowry Definition

Brønsted and Lowry independently defined an acid as a proton donor. The classical acids can also be considered Brønsted acids. Upon ionization the Brønsted acids yield protons and conjugate bases. Conversely, Brønsted bases accept protons to form conjugate acids.

It is important to note that Brønsted acids include both uncharged and ionic species, e.g., HCl, $HSO_4^{\ominus}$, and $NH_4^{\oplus}$, because of their tendency to lose a proton.

1.1.2. The Lewis Definition

The Lewis system defines acids as electron acceptors and bases as electron donors. An acid–base reaction involves the transfer of an electron pair from the base to the acid. Although the Lewis bases are largely the same as the Brønsted bases, the range of acids differs markedly. For example, in the Brønsted system the acids donate a proton, but in the Lewis sense the proton is

1

the acid. The most important generalization of the Lewis theory is that many aprotic compounds are included in the acid category.

Cationic entities of a nonmetallic nature, either stable or transient, are important intermediates in organic reactions. They belong to the class of Lewis acids. Organic groups with polarizable multiple bonds such as carbonyl, cyano, and nitroso groups readily undergo addition with nucleophiles (Lewis bases). These are generally regarded as Lewis acids.

Although the differences between the Brønsted and Lewis bases are minimal compared with that between the acids, Brønsted bases constitute, in fact, a small group of Lewis bases. Besides anions, Lewis bases consist of molecules with lone-pair electrons as well as π-electron systems. π Complexes play significant roles in organic and organometallic chemistry. Many of these can be detected spectroscopically or can be isolated.

1.2. THE HARD AND SOFT ACIDS AND BASES (HSAB) CONCEPT: HISTORICAL DEVELOPMENT

R. G. Pearson has given a lucid account of the historical background concerning the development of the hard and soft acids and bases concept in the Introduction of his book (1) of collected papers.

The collaboration of Pearson and Edwards in 1961 in an incisive evaluation of factors determining nucleophilic reactivities (2) sowed the seed which eventually blossomed into a series of classic papers systematizing chemical stabilities, specificities, and reaction rates (3–5) on the basis of the "Hard and Soft Acid and Bases Principle."

The principle was first applied to inorganic coordination compounds and then used to interpret the chemistry of organic molecules and even electrode (6) and catalytic (7) phenomena. It appears that the principle can be employed in every branch of descriptive chemistry.

Pearson envisages a chemical bond as being made up of an acid–base combination. One can always imaginarily dissect a compound into Lewis acid and Lewis base moieties for analysis. The properties of the intact molecule, including the ease of its formation, can be inferred by considering its acid and base components. The classes of acids and bases are named "hard" and "soft"; the A or (a) metal ions belong to hard acids and the B or (b) metal ions (8, 9) are soft acids.

According to Pearson, the choice of these terms was influenced by S. Winstein's remark that the iodide ion is a "soft, mushy base," and D. Busch's description that the hydroxide ion is a "hard, tight base." Despite occasional scorn from some chemists, the terms have been perpetuated since Pearson's first paper (3) appeared in 1963. In that article, Pearson propounded that *"hard acids bind strongly to hard bases and soft acids bind strongly to soft bases."* This has become known as the *HSAB principle.*

The concept received immediate and enthusiastic acclaim from some chemists and an international symposium on HSAB (pronounced "hassab") was sponsored by Cyanamid European Research Institute in Geneva in May 1965. Two years later, the second symposium in London attracted a greater number of scientists.

In 1967, another major paper by Pearson and Songstad (5), applying the HSAB principle to organic chemistry, appeared. Saville published a full account of his analysis of multicenter reactions (10) based on the HSAB principle. Efforts have been made to determine quantitatively the hardness (softness) of some donors and acceptors (11). Klopman presented a perturbation treatment (12) of chemical reactivity and related HSAB to molecular orbital theory.

Summaries of the HSAB principle have been given by Pearson (13–16) and reviews dealing mainly with organic topics have been written in French (17) and English (18). On the basis of the steadily increasing quotations of Pearson's papers, it may be concluded that the HSAB principle has gained importance in the thought of chemists.

REFERENCES

1. R. G. Pearson, "Hard and Soft Acids and Bases." Dowden, Hutchinson, & Ross, Inc., Stroudsburg, Pennsylvania, 1973.
2. J. O. Edwards and R. G. Pearson, *J. Am. Chem. Soc.* **84,** 16 (1962).
3. R. G. Pearson, *J. Am. Chem. Soc.* **85,** 3533 (1963).
4. R. G. Pearson and J. Songstad, *J. Am. Chem. Soc.* **89,** 1827 (1967).
5. R. G. Pearson and J. Songstad, *J. Org. Chem.* **32,** 2899 (1967).
6. D. J. Barclay, *J. Electroanal. Chem.* **19,** 318 (1968); D. J. Barclay and J. Caja, *Croat. Chem. Acta* **43,** 221 (1971).
7. R. Ugo, *Chim. Ind. (Milan)* **51,** 1319 (1969).
8. G. Schwarzenbach, *Adv. Inorg. Chem. Radiochem.* **3,** 257 (1961).
9. S. Ahrland, J. Chatt, and N. R. Davies, *Q. Rev., Chem. Soc.* **11,** 265 (1958).
10. B. Saville, *Angew. Chem. Inter. Ed. Engl.* **6,** 928 (1967).
11. A. Yingst and D. H. McDaniel, *Inorg. Chem.* **6,** 1067 (1967); S. Ahrland, *Chem. Phys. Lett.* **2,** 303 (1968).
12. G. Klopman, *J. Am. Chem. Soc.* **90,** 223 (1968).
13. R. G. Pearson, *Science* **151,** 172 (1966).
14. R. G. Pearson, *Chem. Br.* **3,** 103 (1967).
15. R. G. Pearson, *J. Chem. Educ.* **45,** 581 and 643 (1968).
16. R. G. Pearson, *Surv. Prog. Chem.* **5,** 1–52 (1969).
17. J. Seyden-Penne, *Bull. Soc. Chim. Fr.* p. 3871 (1968).
18. T.-L. Ho, *Chem. Rev.* **75,** 1 (1975).

2

Classification of Hard and Soft Acids and Bases

2.1. PEARSON'S GENERALIZATION

After collaborating with Edwards to analyze and systematize the rate data (1) for nucleophilic displacement, Pearson focused his attention on a similar treatment of the equilibrium constants (2) for the reaction

$$N + S-X \longrightarrow N-S + X$$

It was concluded that there are two kinds of substrate acids. Members of one class bind strongly to bases which have a high affinity for protons, while those of the second class bind preferentially with highly polarizable bases. The bases are called "hard" and "soft," respectively, as suggested to Pearson by D. H. Busch. The acids with which these bases combine preferentially are the hard and the soft acids. Metal ions of class A (3) or (*a*) (4) are hard acids, and the class B or (*b*) ions are soft.

The hardness of an acid (acceptor) or a base (donor) is generally characterized by a small atomic radius, a high effective nuclear charge, and a low polarizability, whereas softness implies all the opposite properties. Furthermore, the softness of a base can be associated with low electronegativity, easy oxidizability, or empty low-lying orbitals.

The main criterion adopted by Pearson in his classification of Lewis acids is more or less the same as that employed by Ahrland *et al.* (4). Hard acids will form complexes whose stability sequences are

$$N \gg P > As > Sb > Bi$$
$$O \gg S > Se > Te$$
$$F \gg Cl > Br > I$$

Conversely, the complex stability of soft acids with bases decreases as the base varies according to the following pattern.

$$N \ll P > As > Sb > Bi$$
$$O \ll S < Se \sim Te$$
$$F \ll Cl < Br < I$$

A number of Lewis acids have been assigned to hard or soft groups according to this general scheme (see Table 2.1). Since no one can determine the demarcation line between a hard and a soft species, there exist certain intermediate or borderline cases. In fact, the hardness (or softness) of a central atom is influenced by its ligands.

TABLE 2.1   Classification of Hard and Soft Acids[a]

Hard acids

H^+, Li^+, Na^+, K^+ (Rb^+, Cs^+)
Be^{2+}, $Be(CH_3)_2$, Mg^{2+}, Ca^{2+}, Sr^{2+} (Ba^{2+})
Sc^{3+}, La^{3+}, Ce^{4+}, Gd^{3+}, Lu^{3+}, Th^{4+}, U^{4+}, UO_2^{2+}, Pu^{4+}
Ti^{4+}, Zr^{4+}, Hf^{4+}, VO^{2+}, Cr^{3+}, Cr^{6+}, MoO^{3+}, WO^{4+}, Mn^{2+}, Mn^{7+}, Fe^{3+}, Co^{3+}
BF_3, BCl_3, $B(OR)_3$, Al^{3+}, $Al(CH_3)_3$, $AlCl_3$, AlH_3, Ga^{3+}, In^{3+}
CO_2, RCO^+, NC^+, Si^{4+}, Sn^{4+}, CH_3Sn^{3+}, $(CH_3)_2Sn^{2+}$
N^{3+}, RPO_2^+, $ROPO_2^+$, As^{3+}
SO_3, RSO_2^+, $ROSO_2^+$
Cl^{3+}, Cl^{7+}, I^{5+}, I^{7+}
HX (hydrogen-bonding molecules)

Borderline acids

Fe^{2+}, Co^{2+}, Ni^{2+}, Cu^{2+}, Zn^{2+}
Rh^{3+}, Ir^{3+}, Ru^{3+}, Os^{2+}
$B(CH_3)_3$, GAH_3
R_3C^+, $C_6H_5^+$, Sn^{2+}, Pb^{2+}
NO^+, Sb^{3+}, Bi^{3+}
SO_2

Soft acids

$Co(CN)_5^{3-}$, Pd^{2+}, Pt^{2+}, Pt^{4+}
Cu^+, Ag^+, Au^+, Cd^{2+}, Hg^+, Hg^{2+}, CH_3Hg^+
BH_3, $Ca(CH_3)_3$, $GaCl_3$, $GaBr_3$, GaI_3, Tl^+, $Tl(CH_3)_3$
CH_2, carbenes
π acceptors: trinitrobenzene, chloroanil, quinones, tetracyanoethylene, etc.
HO^+, RO^+, RS^+, RSe^+, Te^{4+}, RTe^+
Br_2, Br^+, I_2, I^+, ICN, etc.
O, Cl, Br, I, N, $RO\cdot$, $RO_2\cdot$
M^0 (metal atoms) and bulk metals

[a] Reprinted with permission from *J. Chem. Educ.* **45**, 581 (1968). Copyright by the American Chemical Society.

The softness of a base is defined by the equilibrium

$$MeHg^{\oplus}(aq) + BH^{\oplus}(aq) \rightleftharpoons MeHgB^{\oplus}(aq) + H^{\oplus}(aq)$$

If the equilibrium constant is much greater than unity, the base B is soft. If it is near unity or less than unity, the base is considered hard. Using this standard, common bases are classified as shown in Table 2.2 (5).

TABLE 2.2 Classification of Hard and Soft Bases[a]

Hard bases

NH_3, RNH_2, N_2H_4
H_2O, OH^-, O^{2-}, ROH, RO^-, R_2O
CH_3COO^-, CO_3^{2-}, NO_3^-, PO_4^{3-}, SO_4^{2-}, ClO_4^-
F^- (Cl^-)

Borderline bases

$C_6H_5NH_2$, C_5H_5N, N_3^-, N_2
NO_2^-, SO_3^{2-}
Br^-

Soft bases

H^-
R^-, C_2H_4, C_6H_6, CN^-, RNC, CO
SCN^-, R_3P, $(RO)_3P$, R_3As
R_2S, RSH, RS^-, $S_2O_3^{2-}$
I^-

[a] Reprinted with permission from *J. Chem. Educ.* **45**, 581 (1968). Copyright by the American Chemical Society.

It should be noted that not only ordinary stable cationic, anionic, and molecular (neutral) Lewis acids and bases are included, but also highly reactive intermediates such as carbenes, free radicals, and hypothetical species.

A generalized statement concerning the behavior of these acids and bases can be made: *Hard acids prefer to bind to hard bases and soft acids prefer to bind to soft bases.* This statement, known as the *Hard and Soft Acids and Bases Principle*, is an extremely powerful qualitative rule in correlating chemical information.

It has been recognized for a long time that a strong acid forms a strong bond with a strong base, i.e., the equilibrium constant of the reaction

$$A + :B \rightleftharpoons A:B$$

is determined by the strength factor S. However, such a simple correlation is not always adequate and a four-parameter equation appears to be the minimum requirement. Pearson proposed that the equilibrium be expressed by

$$\log K = S_A S_B + \sigma_A \sigma_B$$

where σ_A, σ_B are measures of some characteristics different from that of strength and are called "softness." The equation is similar to that of Edwards (6).

$$\log (K/K_0) = \alpha E_n + \beta H$$

The α and β values for many common metal ions have been determined (7). For Lewis acids with a high positive charge and small size, β is large, whereas Lewis acids with low charge and large size have small β values. It follows that β corresponds to Pearson's S_A; therefore, $\beta H = S_A S_B$ and $\alpha E_n = \sigma_A \sigma_B$. The softness parameters are correlated with the oxidation potential and, hence, the polarizability of the acids and bases.

2.2. SOFTNESS SCALES

Much effort has been directed toward finding a universal softness parameter σ. Pearson and Mawby (8) calculated the coordinate bond energy (CBE) for metal halides and defined a parameter $[CBE(F^-) - CBE(I^-)]/CBE(F^-)$ which provides a measure of the softness of the acceptor (acid) within a charge group (e.g., $M^{\oplus}$, $M^{2\oplus}$, $M^{3\oplus}$, etc).

Klopman's theoretical treatment (9) of HSAB leads to the conclusion that the hardness of an acceptor is associated with the low energy of its empty frontier orbital relative to desolvation energy of the acceptor. He also was able to estimate the energy difference, with reference to water. Very hard acceptors have large negative values and very soft ones have large positive values; generally they are charge-independent. Softness parameters for donors (bases) have been similarly calculated.

Ahrland (10) reasoned that the more completely the energy for positive ion formation in the gas phase is regained by introduction of the ion in a hard solvent (e.g., H_2O) the harder the ion is. He then proposed a softness parameter based on the dehydration energy and the ionization potential for the formation of $M^{n\oplus}$(gas). A large difference between the two quantities indicates a soft ion.

Yingst and McDaniel (7) used the α/β ratio from Edwards equation as a criterion for assessing the softness of metal ions. It is shown that hard acids generally have a low α (polarizability) and a high β (basicity) value, therefore α/β permits simultaneous consideration of both factors.

Despite a certain degree of success achieved from these and other attempts, exceptions have always been noted. Pearson himself has questioned (11) the overall usefulness of quantification of the HSAB concept. Apart from the inherent limitations of each approach, it is difficult to imagine that prescribed values are flexible enough to give a meaningful picture when dealing with external influences such as ligand effects and ambient behavior.

Although an exact softness scale is neither available nor advisable, empirical systematization of various acids and bases is essential for correlating experimental observations. The best compromise would be to establish guidelines for the rough estimation of softness in a comparative sense.

A general consensus is that hardness increases (or softness decreases) with increasing positive oxidation state. For example, Ni(0) is soft, Ni(II) is borderline, but Ni(IV) is hard; the sulfur atom in $RS^{\ominus}$ is soft, medium soft in $RSO^{\ominus}$, and it is hard in $RSO_2^{\ominus}$; $S^{2\ominus}$ is softer than $SO_3^{2\ominus}$. There are a few exceptions to the rule, however. For instance, Tl(III), Sn(IV), and Pb(IV) are softer than their respective lower valent ions. Because Tl(I), Sn(II), and Pb(II) ions have $d^{10}s^2$ electrons in their outermost shells, the shielding of the d electrons decreases the softness of the lower valent species (12). One factual demonstration of this reverse hardness/valence relationship is that inorganic thallium compounds are generally more stable in the +1 valence state, while covalent organothallium derivatives are stable only in the +3 state.

A special case of the nonmonotonic relationship of hardness with oxidation number has been discussed by Jørgensen (13). Manganese ion changes from soft to hard and then reverts to soft behavior as a function of its oxidation state.

It may be that hardness correlates with fractional charge on the central atom rather than its oxidation number. This rationalization receives support from observations of ligand effects. Despite a formal $B^{3\oplus}$ core possessed by both BF_3 and BH_3, the former compound is a hard Lewis acid, and the latter is a soft one. This occurs because hydride ions effectively reduce the charge of the boron atom in borane. This phenomenon is particularly evident with soft donor ligands from which the negative charge is easily transferred.

For a series of bases with congeneric central atoms, the heavier or more electropositive member is usually the softer: $R_3P > R_3N$; $R_2Se > R_2S > R_2O$; $I^{\ominus} > Br^{\ominus} > Cl^{\ominus} > F^{\ominus}$. The isoelectronic anions from the lower family on the same row (period) of elements are the softer: $CH_3^{\ominus} > NH_2^{\ominus} > OH^{\ominus} > F^{\ominus}$.

Since strength and softness are two independent properties, a base can be both soft and strongly binding toward protons. Nonstabilized carbanions are cases in point.

The softness of a carbanion is related to the hybridization of the carbon. Higher p character increases the softness ($sp^3 > sp^2 > sp$) and configurational instability of the carbanion. Reactions involving carbanions formally derived from alkanes often lead to racemization, whereas configuration is retained at the alkenic carbanion center. The fact that cyclopropyl carbanions maintain their original stereochemistry (14) concurs with the above notion, as the s character of cyclopropane exocyclic orbital is $> 25\%$. Surprisingly, even an α-cyano group on the ring does not accelerate racemization (15) to a great extent.

Carbon acids are relatively soft. Owing to the electron-withdrawing property of alkyl groups attached to a sp^3-hybridized carbon (16), the hardness sequence of several carbenium ions follows the order $Ph^\oplus > t\text{-}Bu^\oplus > i\text{-}Pr^\oplus > Et^\oplus > Me^\oplus$. This sequence is supported by thermodynamic data deduced from the reactions of alcohols with hydrogen sulfide. The validity of the above sequence need not be restricted to free cations,* but it implies that the more carbenium character a center attains during a reaction, the harder it will be. The increasing stability of isomeric butanols ($n < iso < sec < tert$) parallels the trend of hardness exhibited by the carbenium ion $R^\oplus$ which combines with the hard hydroxide ion. This offers an explanation for isomerizations such as

$$i\text{-BuOH(g)} \longrightarrow t\text{-BuOH(g)}$$

Replacement of the hydrogen atoms of $CH_3^\oplus$ by electronegative groups would certainly harden the cation. Since $H^\ominus$ is one of the softest bases, $CH_3^\oplus$ represents the extreme case on the softness scale of the carbon acids bearing a positive charge. The only way to improve on its softness is to remove a proton thereby creating a carbene $:CH_2$. A carbon free radical is soft both as an acceptor and as a donor. Olefins act as soft bases by using the π-electron system.

2.3. THEORETICAL DESCRIPTIONS OF HSAB

The reason for the wide applicability and validity of the HSAB principle must be of a very fundamental nature, and the simplest and most apparent basis is the ionic–covalent dualism accounting for the hard–hard and soft–soft interactions. Ionic bonding is favored by small size and large charge which are exactly the properties possessed by hard acids and bases. On the other hand, good covalent bonding requires partners of similar size and electronegativity. As many soft acids do not have any charge at all, it can hardly be anticipated that they could participate in ionic bonding.

The mutual affinity of soft acids and bases could be attributed, at least in part, to London or van der Waals dispersion forces which depend on the products of polarizabilities, and these forces are large when both interacting groups are highly polarizable.

Chatt (17) has proposed a π-bonding theory for metal ions. Class (b) acids

---

* Free ions exist only in the gas phase. In our discussion these are only formal representations.

have loosely held outer d-orbital electrons which can form π bonds by back-donation to suitable ligands.

Mulliken (18) considered the extra stability of the bonds between large atoms as being derived from d–p hybridization, whereby some d character is instilled into both π and π^* molecular orbitals. This has the effect of increasing bonding orbital overlap, but decreasing the overlap of antibonding orbitals. The result could be dramatic if large atoms with deep mutual penetration of the electron clouds are involved as repulsion, according to the Pauli principle, is greatly alleviated.

By far the most thorough theoretical treatment of the HSAB relationship is that of Klopman (9). He applied the quantum mechanical perturbation method to analyze the events following the interaction of two systems R and S. The total perturbation energy from such an interaction is considered as being generated by a combination of (1) neighboring effects which accounts for interaction due to ion-pair formation without electron transfer and (2) partial charge transfer during covalent bonding. The most important feature of this approach is the inclusion of solvation phenomena.

When the perturbation under consideration is small, the total perturbation energy can be expressed as

$$\Delta E_{total} = -q_R q_S \frac{\Gamma}{\epsilon} + \Delta solv(1) + \sum_{\substack{m \\ occ}} \sum_{\substack{n \\ unocc}} \left[\frac{2(c_R^m)^2 (c_S^n)^2 \beta^2}{E_m^* - E_n^*} \right]$$

where q_R and q_S are total initial charges of atoms R and S, respectively, Γ is the Coulomb repulsion term between R and S, ϵ is the local dielectric constant of the solvent, E_m^* is the highest occupied orbital of the donor, and E_n^* is the lowest unoccupied orbital of the acceptor.

When $|E_m^* - E_n^*| \gg 4\beta^2$, very little charge transfer occurs. Thus, ΔE_{total} becomes

$$-q_R q_S \frac{\Gamma}{\epsilon} + \Delta solv(1) + 2 \sum_{\substack{m \\ occ}} (c_R^m)^2 \sum_{\substack{n \\ unocc}} (c_S^n)^2 \gamma$$

where $\gamma = \beta^2 / (E_m^* - E_n^*)_{av}$.

The total charges on the two reactants are the dominant factors of the reaction (charge-controlled reaction). This type of reaction occurs when the donor is difficult to ionize or polarize (E_m^* very low) and the acceptor has a slight tendency to accept electrons (E_n^* very high) and when both are strongly solvated, i.e., they are small. The charge-controlled reaction is therefore synonymous with a hard–hard interaction.

When $|E_m^* - E_n^*| \approx 0$, interaction of the frontier orbitals becomes important, and a strong electron transfer takes place (frontier-controlled reaction). The reactivity

$$\Delta E = 2c_R^m c_S^n \beta$$

is determined by the frontier electron density (c_R^m, c_S^n) when uncharged or weakly charged species interact. Polarizability of the reactants and low solvation energies occur in the reaction, therefore such an interaction is identified with soft–soft interactions. Table 2.3 summarizes both types of reactions. Only the hard–hard and soft–soft interactions lead to a high reactivity.

TABLE 2.3   Type and Rate of Reaction between Hard and
Soft Reagents[a]

Donor E_m^*	Acceptor E_n^*	$E_m^* - E_n^*$	Γ	β	Reactivity	
High (soft) large orbital	High (hard) small orbital	Medium	Small	Very small	Undefined	Low
	Low (soft) large orbital	Small	Very small	Large	Frontier controlled	High
Low (hard) small orbital	High (hard) small orbital	Large	Large	Small	Charge controlled	High
	Low (soft) large orbital	Medium	Small	Very small	Undefined	Low

[a] Reprinted with permission from Klopman (9). Copyright by the American Chemical Society.

Klopman has succeeded in calculating the softness character of many cations and anions. The correlation to experimental observations is surprisingly good.

REFERENCES

1. J. O. Edwards and R. G. Pearson, *J. Am. Chem. Soc.* **84**, 16 (1962).
2. R. G. Pearson, *J. Am. Chem. Soc.* **85**, 3533 (1963).
3. G. Schwarzenbach, *Adv. Inorg. Chem. Radiochem.* **3**, 257 (1961).
4. S. Ahrland, J. Chatt, and N. R. Davies, *Q. Rev., Chem. Soc.* **11**, 265 (1958).
5. R. G. Pearson and J. Songstad, *J. Am. Chem. Soc.* **89**, 1827 (1967).
6. J. O. Edwards, *J. Am. Chem. Soc.* **76**, 1540 (1954).
7. A. Yingst and D. H. McDaniel, *Inorg. Chem.* **6**, 1067 (1967).
8. R. G. Pearson and R. J. Mawby, *in* "Halogen Chemistry" (V. Gutmann, ed.), Vol. 3, p. 55. Academic Press, New York, 1967.
9. G. Klopman, *J. Am. Chem. Soc.* **90**, 223 (1968).

10. S. Ahrland, *Chem. Phys. Lett.* **2**, 303 (1968).
11. R. G. Pearson, "Hard and Soft Acids and Bases," p. 242. Dowden, Hutchinson, & Ross, Inc., Stroudsburg, Pennsylvania, 1973.
12. S. Ahrland, *Struct. Bonding (Berlin)* **1**, 207 (1966).
13. C. K. Jørgensen, *Struct. Bonding (Berlin)* **1**, 234 (1966).
14. H. M. Walborsky, F. J. Impastato, and A. E. Young, *J. Am. Chem. Soc.* **86**, 3283 (1964).
15. H. M. Walborsky, *Rec. Chem. Prog.* **23**, 75 (1962).
16. R. C. Fort, Jr., and P. v. R. Schleyer, *J. Am. Chem. Soc.* **86**, 4194 (1964); V. W. Laurie and J. S. Muenter, *ibid.* **88**, 2883 (1966).
17. J. Chatt, *J. Inorg. Nucl. Chem.* **8**, 515 (1958).
18. R. S. Mulliken, *J. Am. Chem. Soc.* **77**, 885 (1955).

3

Chemical Reactivity

3.1. STABILITY OF ORGANIC COMPOUNDS AND COMPLEXES

In discussing the stability of compounds, a frame of reference should always be provided. A compound that is stable under one set of conditions can become very labile under others. Moreover, stability could be referred to thermodynamically or kinetically. In assessing the relative stability of organic species, the HSAB principle is extremely helpful. However, before proceeding further, one should remember that proper consideration of the partners in an acid–base complex is paramount to successful application (1).

The acyl group $RCO^{\oplus}$ is a hard Lewis acid, hence its combination with hard bases forms thermodynamically stable molecules, e.g., carboxylic acids RCOOH, esters RCOOR', and amides $RCONR'_2$. In contrast, its union with soft bases results in highly reactive or labile species such as thioesters RCOSR', selenoesters RCOSeR', and acyl iodides RCOI.

An oxy group, when acting as an acceptor (e.g., $RO^{\oplus}$), is soft. Therefore, it is easily understood why peroxycarboxylic acids are thermodynamically un-stable and reactive molecules. Moreover, the stronger (harder) the carboxylic acid, the more reactive its corresponding peracid is since there exists a wider "softness gap" between $OH^{\oplus}$ and $RCOO^{\ominus}$. Thus, trifluoroperacetic acid is a more potent oxidant than peracetic acid, and triazole-1-peroxycarboxylic acid (2) is about two hundred times more reactive than perbenzoic acid toward olefins.

By the same token, acyloxyhalides RCOOX are extremely labile by virtue of the hard $(RCOO^{\ominus})$–soft $(X^{\oplus})$ combination. They exist as transient inter-mediates in the Hunsdiecker and Simonini reactions (3).

$$RCOO^{\ominus} Ag^{\oplus} + X_2 \xrightarrow[-AgX]{} RCOOX \longrightarrow RX + CO_2$$

Organosilyl perbenzoates cannot be isolated as they undergo rearrangement much more readily than the carbon analogs (4). The bonding situation in the percarboxylates forces the oxygen adjoining the hard silicon atom into a role of soft acceptor. The incompatibility of the donor and acceptor moieties of these molecules with respect to their hardness contributes to their lability. Furthermore, an excellent opportunity for rearrangement to the more stable isomers presents itself (5).

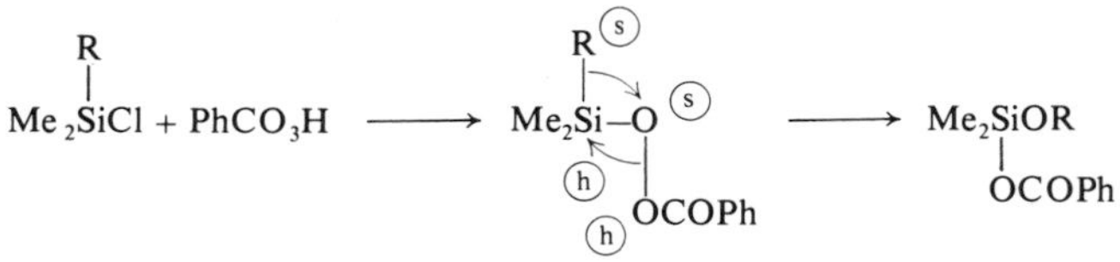

Strong transannular N· · ·C=O interaction in the mesocycle (1) (6, 7) has been shown. The corresponding sulfur analogs (8) show weaker interactions. When the hetero group is a sulfoxide, participation by its oxygen is observed (9) (see 3). The sulfoxide ammonium salt (4) assumes a conformation (10) in which hydrogen bonding between NH and the sulfoxide oxygen prevails. Thus, predilection for hard–hard interaction is clear. Additional examples are found in the transannularly bridged salt formation from 1-thiacycloheptan-4-one (11) and 3α-phenyl-3β-tropanyl phenyl ketone (12).

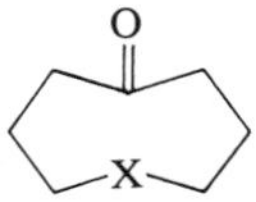

(1) 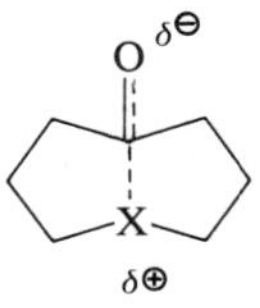　　　　　　(2) 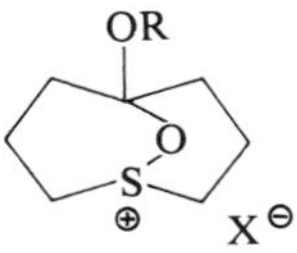　　　　　　(3) 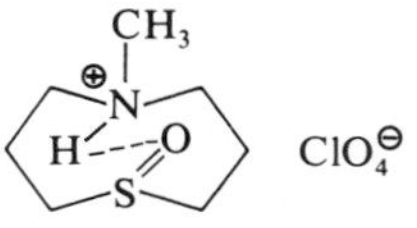　　　　　　(4)

In 2-halocyclohexanones there is a shift toward axial preference for the halogen (13) as it becomes more polarizable. The $O:X_{eq}$ interaction would be more highly destabilizing if the hardness difference of the heteroatoms is more pronounced.

Lewis acid coordination with carbonyl compounds lowers the C=O stretching frequency in the infrared. Because the donor atom is the hard oxygen atom, tighter and more stable complexes are formed between carbonyl compounds and hard Lewis acids, and these species are detected by infrared spectroscopy (14). The following shifts exhibited by acetophenone are typical.

	$FeCl_3$	$AlCl_3$	$TiCl_4$	BF_3	$ZnCl_2$	$CdCl_2$	$HgCl_2$
$\Delta\nu_{CO}$ (cm^{-1})	130	120	118	107	47	38	31

The substituent effect on the enolization of carbonyl compounds may be analyzed in the HSAB context. When X of (5) is hydrogen, C-1 is softened and enolization is favored. The C-1 of the enol form (6) is probably softer since it is doubly bonded to another carbon instead of to oxygen as in the keto form. Aldol condensation (15) proceeds much more readily with aldehydes than with ketones, and enolization almost invariably preludes C–C bond formation (16). When X = Cl, OH, OR, and other electronegative groups, C-1 becomes harder and enolization is thereby discouraged.

α-Substitutents (Z) affect the ease of enolization accordingly. Electronegative groups such as halogen or carbonyl prefer direct bonding to a harder center which is available in the enol form. Facilitation of enolization by these substituents is well-recognized. The enol content of several 2-substituted ethyl acetoacetates (17) increases with the hardness of the substituent, except for those capable of π–π conjugation, as noted for the following substituents (% enol): H(8), Me(5), Et($\sim$1), Ph(30), F(15), Cl(15), Br(5), CF_3(89), and CN(93).

(5) (6)

Organic nitriles form adducts with hydrogen halides. The thermodynamic stability of these imidyl halide salts (18–20) increases with the acidity of HX. The fact that the nitrile–hydrogen fluoride adducts are particularly stable (21) reflects the favorable interaction of hard bases with the hard nitrile carbon.

$$RCN + 2HX \longrightarrow$$

X = Cl, Br, I

Thermal reorganization of isonitriles to nitriles (22) can be expected on the basis of the HSAB concept, although isomerization of some compounds proceeds via a free radical mechanism (23, 24).

Equilibration of thiocyanates yields a partition ratio of $k_S/k_N \simeq 5$, indicating that the thiocyanate is more stable than the isothiocyanate (25.). The soft sulfur atom prefers bonding singly rather than doubly to the carbon.

$$RSCN \rightleftharpoons R^\oplus \ SCN^\ominus \rightleftharpoons RNCS$$
(79%)   (16%)

Thiocyanates form more stable complexes with phenol (through hydrogen bonding) than with iodine (by charge transfer), whereas exactly opposite results are observed with isothiocyanates (26).

The strength of the interaction between phenol and dialkyl chalcogenides (R_2X) falls off as $X = O \gg S > Se$ (27), and between phenol and alkyl halides as $RF > RCl > RBr > RI$ (28). These findings are in keeping with the relative compatibility in hardness of the proton and the hydrogen bond acceptor. 2-Pyridones and the corresponding thiones and selenones are highly polarized. In fact, the dipole moment increases along the series, but the hydrogen bonding ability decreases progressively (29).

The stability of silver complexes (AgL, $AgHL^{\oplus}$, $AgL_2^{\ominus}$) with $ArXCH_2COO^{\ominus}$-type ligands (L) has been measured (30), and found to be a function of X ($X = Se > S > O$).

The ease of complex formation decreases along the series $Me_2O > Me_2S > Me_2Se$ with BF_3 as reference acceptor, but with BH_3 and BMe_3 an almost reverse trend is observed ($Me_2S > Me_2Se \sim Me_2O$). For $X = S$ or Se, $Me_2X \cdot BH_3 > Me_2X \cdot BF_3$ (31). BX_3 coordinates better with amines than with ethers and sulfides (32).

From these studies it may be concluded that boron is a borderline acceptor which becomes hard when surrounded by hard ligands, and it is rendered soft by soft substituents (e.g., $H^{\ominus}$). 3-(Methylthio)propylborane (7) is more stable than the borane–tetrahydrofuran complex (8) and it is distillable (33). The boron atom of the dihydroborepin (9) is internally coordinated by the amino nitrogen; however, this chelation is removed by the addition of trityl fluoroborate (34). This may indicate that the boron atom in the borinic esters (10) is harder, but not as hard as the trityl cation.

(7)   (8)   (9)   (10)

Equilibration studies (35, 36) yield data concerning hard–soft characteristics of donors and acceptors.

$$Me_3NBH_3 + Me_3P \xrightleftharpoons{\quad 20:80 \quad} Me_3PBH_3 + Me_3N$$

$$Me_2PNMe_2 + Me_2PH \xrightleftharpoons{\quad 11:89 \quad} Me_2PPMe_2 + Me_2NH$$

The relative inertness of divalent sulfur and trivalent phosphorus compounds toward proton and their avidity in reaction with soft alkyl halides are well documented. Competition for alkyl halides among nitrogen and phosphorus atoms in aminophosphines (37, 38) serves to illustrate the usefulness of the HSAB principle.

$$R_2NPR_2' + R''X \longrightarrow R_2NP^{\oplus}R_2'R'' \ X^{\ominus}$$

The stability sequences of tetrahalosilane–trimethylamine and tetrahalosilane–phosphine adducts have been determined (39). Silanes containing harder halogens prefer complexation with the harder base.

Alkyl hypohalites are very reactive molecules from which positive halogens are easily liberated when attacked by nucleophiles. The instability and reactivity (ROI > ROBr > ROCl) can be traced to the hard–soft mixing of adjoining heteroatoms. The fact the triorganosilyl hypohalites are even more unstable (40) is consistent with the HSAB principle, as the oxygen in these compounds is further hardened by the neighboring silicon resulting in a more pronounced HSAB differential with respect to the soft halogen acceptor.

The sulfenate esters RSOR' also have a soft–hard make-up; they are more reactive and less stable than disulfides RSSR', in which both the acid and base moieties are soft.

Evidence has been obtained for the following equilibria (41) which involve soft–soft acid–base interactions.

$$Me_2S^{\oplus}-SMe \ BF_4^{\ominus} + Me_2S_2 \ \rightleftharpoons \ Me_2S + (Me_2S)_2S^{\oplus}Me \ BF_4^{\ominus}$$

$$(CH_3)_2S^{\oplus}SCH_3 \ BF_4^{\ominus} + Me_2S \ \rightleftharpoons \ (CH_3)_2S + Me_2S^{\oplus}SCH_3 \ BF_4^{\ominus}$$

Mixed soft–hard interactions are so unfavorable in the equilibria portrayed below that either they are biased completely to one side or they are immeasurably slow by the nmr time scale.

$$Me_2S^{\oplus}SMe + Me_2O \ \overset{\times}{\rightleftharpoons} \ Me_2S + Me_2O^{\oplus}SMe$$

$$(CH_3)_2S^{\oplus}OCH_3 + Me_2S \ \overset{slow}{\rightleftharpoons} \ (CH_3)_2S + CH_3OS^{\oplus}Me_2$$

Compounds $R_3M{=}X$ (M = P, As; X = S, Se) are excellent ligands for soft metal ions, but not for hard acceptors (42). This behavior is at variance with compounds $R_3M{=}O$ (M = P, As) which display metal ion affinity in the opposite sense.

The lanthanide cations are hard Lewis acids. Specific coordination of organic functional groups with complexed salts of this subgroup of elements forms the basis of lanthanide-induced shifts observed in nuclear magnetic resonance of organic molecules (43–45). Larger shifts of the nearby hydrogen atoms are associated with tighter coordination of the functional group with the metal ion, hence an approximate gauge for determining the relative hardness of heteroatom bases in different environments is available.

Larger shifts in carbon-13 nmr of carbonyl compounds are induced by titanium tetrachloride (46) compared to those observed in the presence of Eu-(fod)₃[tris(6,6,7,7,8,8,8-heptafluoro-2,2-dimethyl-3,5-octanedionato)europium]. Ti(IV) is harder than Eu(III).

σ-Bonded organometallic species can be considered as $M^⊕ R^⊖$ acid–base pairs. Since $R^⊖$ is a soft base, it is reasonable to assume that organometallic compounds are more stable if $M^⊕$ is soft. Confirming examples are too extensive to enumerate, but it may be mentioned that Grignard reagents are more stable than the corresponding organolithium compounds (RLi) which, in turn, are more stable than RNa and RK. Organomercury, thallium and lead derivatives are especially stable.

Most metal ions form 2,4-pentanedione chelates having the general structure (11). It is of particular interest to note that the mercuric ion is σ-bonded to the central carbon of the β-diketones (47). The softness of Hg(II) must be taken into account in this case.

(11)   (12)

Isodesmic hydride exchanges (48) show that the primary carbocations $MeHgCH_2CH_2^⊕$ and $Me_3SnCH_2CH_2^⊕$ are kinetically more stable than trityl cation (49). These species can be considered as three-centered bound ethylene metal ions (50) in which a considerable amount of the positive charge is carried by the metal. Only the relatively soft metals can be expected to interact so strongly with the moderately soft carbenium center.

Many organometallic compounds owe their stability to soft–soft interactions. Zerovalent heavy metals are soft acceptors; therefore, they are usually bonded to soft ligands such as olefins, isonitriles, carbon monoxide, sulfides, and phosphines.

An unusual complex, di-μ-acrylonitrilebis(tricarbonyliron), contains π and n donors (51). The nitrogen occupies an axial position which is preferred by harder ligands.* However, the most striking feature is the nonlinear CN → Fe

---

* The apical preference for harder ligands is further demonstrated by the following sulfurane structures (52, 53).

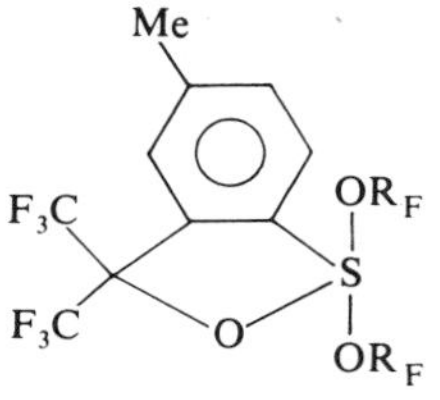

$R_F = C(CF_3)_2Ph$

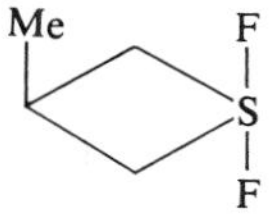

bond. It is possible that because of this orientation the nitrogen donor becomes softer.

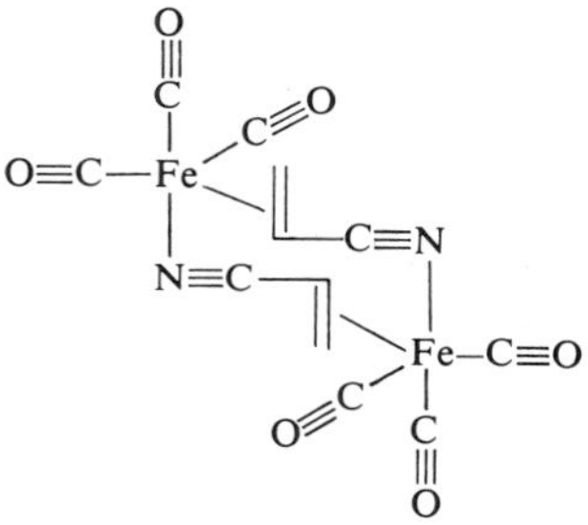

Many stable metal–carbene complexes (54–57) have been synthesized in recent years. All these have soft or borderline central atoms which are further softened by the other soft ligands present. Hydroxycarbene complexes (58) have also been isolated.

When methoxycarbene pentacarbonyl complexes of Cr, Mo, and W are reacted with boron trihalides, loss of the methoxy group and a carbon monoxide ligand results. The metal becomes coordinated to a halogen and a carbyne (59).

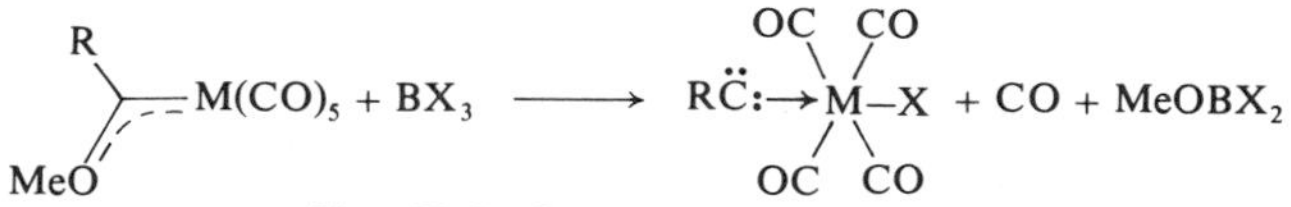

$$X = Cl, Br, I$$

A number of reactions have been postulated as involving metal–carbenoid intermediates.

Diazo compounds lose nitrogen readily on contact with heavy metals (Ag, Cu, Pd) and their salts. These metallic species provide vacant orbitals to accept an electron-pair from the diazo carbon atom and simultaneous back-donation aids in the expulsion of nitrogen (60–62). In other words, the soft–soft interactions throughout the biphilic process (63) account for the effectiveness of the catalysts.

Treatment of diazo compounds with an excess of nickel carbonyl in ethanol leads to the formation of carboxylic esters (64).

$$R_2C^{\ominus}-N_2^{\oplus} + Ni(CO)_4 \xrightarrow[-CO]{} R_2C-Ni^{\ominus}(CO)_3 \xrightarrow[-N_2]{}$$
$$\underset{\oplus N_2}{|}$$

$$R_2C^{\oplus}\!\!-\!\!Ni^{\ominus}(CO)_2 \xrightarrow[-Ni(CO)_2]{} R_2C=C=O \xrightarrow{EtOH} R_2CHCOOEt$$
$$\underset{CO}{|}$$

It is likely that the reduction of organic azides with vanadium(II) chloride involves the formation of vanadium coordinated nitrene intermediates (65). A nitrene coordinated-Ru(II) ion (65a) produced by treatment of Ru(III) azide complexes with acid has been identified.

The phosphorus and sulfur ylides are more stable than the nitrogen and oxygen analogs (66). The former species are comprised of carbenes complexed to soft donors, whereas in the latter the carbenes are not stabilized by the adjoining hard bases. The high stability of $CH_2I^{\ominus}$ compared to that of $CH_2F^{\ominus}$ is at variance with the classical theory of inductive effect, but it is exactly as predicted on the basis of HSAB principle if regarded as $[X^{\ominus} \rightarrow \ddot{C}H_2]$ complexes. The hard $F^{\ominus}$ is not expected to interact strongly with the soft carbenic center. Likewise, the enhanced reactivity of $FC(NO_2)_2^{\ominus}$ (67) may be considered as a consequence of the destabilization of the hypothetical dinitrocarbene by the hard halide ion.

Iodide ion-initiated decomposition of the mercurials (13) proceeds faster with $X = CF_3$ than with $X = F$. The fluorobis(trifluoromethyl)methide ion intermediate is less stable than tris(trifluoromethyl)methide (68). The efficiency of dichlorocarbene interception by various anions has been compared (69) and these fall into the anticipated sequence $I^{\ominus} \sim Br^{\ominus} > Cl^{\ominus} > F^{\ominus}, NO_3^{\ominus}$, $ClO_4^{\ominus}$. However, $F^{\ominus}$ is more reactive than $Cl^{\ominus}$ toward difluorocarbene (70).

$$[(CF_3)_2CX]_2Hg + 2I^{\ominus} \longrightarrow \underset{CF_2}{\overset{F_3C\diagdown\quad\diagup X}{\big\|}} + 2F^{\ominus} + HgI_2$$
$$\text{(13)}$$

Silicon compounds containing a Si–H bond are thermodynamically unstable because $Si^{\oplus}$ and $H^{\ominus}$ are hard and soft, respectively. The thermal isomerizations of these substances are known to occur (71) under conditions where the carbon analogs remain unperturbed, e.g.,

$$ArCH_2SiHMe_2 \xrightarrow{\Delta} ArSiMe_3$$

3.2. SYMBIOSIS

Jørgensen (72) noted that metal ions prefer bonding to ligands of the same kind whether they are hard or soft. Hard–soft mixed ligature results in unstable species. The term "symbiosis" is applied to describe the phenomenon of maximum flocking of either hard or soft ligands in the same complexes.

Although symbiosis may be an improper term according to its biological significance, identification of the phenomenon has tremendous implications in correlating the behavior of chemical entities. Its validity for organic compounds has been established.

The higher affinity of $:CF_2$ for $F^{\ominus}$ than for $Cl^{\ominus}$ as mentioned in the previous section is an example of symbiosis. In this connection Hine pointed out that (73) C–F and C–O bonds in the same molecule tend to reinforce each other.

There is a gradual C–F bond contraction along the series CH_3F (1.39 Å), CH_2F_2 (1.36 Å), CHF_3 (1.33 Å), and CF_4 (1.32 Å). This observation led to the postulation of a "reverse" hyperconjugation involving fluorine (74, 75). The shorter and hence stronger C–F bond is a consequence also predicted by symbiosis.

The fact that BF_3 readily accepts another $F^{\ominus}$ to form $BF_4^{\ominus}$ and BH_3 gains a hydride ion to give $BH_4^{\ominus}$ are well-known.

gem-Difluoro Meisenheimer complex from picryl fluoride (76) has been detected by nmr, whereas evidence for the formation of the corresponding gem-H, F complex is ambiguous.

Disproportionation of difluoromethane and formaldehyde and the halogen exchange between iodotrifluoromethane and fluoromethane are assisted by symbiosis.

$$2CH_2F_2 \longrightarrow CH_4 + CF_4$$
$$2CH_2O \longrightarrow CH_4 + CO_2$$
$$CF_3I + CH_3F \longrightarrow CF_4 + CH_3I$$

For hydrocarbon molecules, symbiosis implies that those containing a maximum number of either C–H bonds (CH_4) or C–C bonds (e.g., Me_4C) are the most stable. This simple rule offers an explanation for the extra stability of highly branched hydrocarbons.

Spontaneous disproportionation of phenyldifluorophosphine (77) occurs at room temperature.

$$2PhPF_2 \longrightarrow \frac{1}{n}(PhP)_n + PhPF_4$$

Decomposition of methyl *o*-phenylenephosphite-*O*-ethyl phenylsulfenate adduct (78) *in situ* must be dictated by symbiosis.

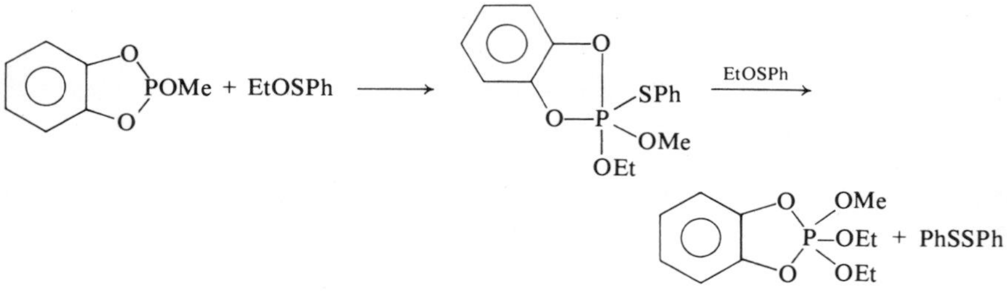

The methylation of *N,N'*-bis(trimethylsilyl)hydrazine (79) occurs with partial rearrangement, which may be interpreted by the symbiotic effect. In the anomalous product each nitrogen atom is bonded twice to the same substituent as well as to one another.

$$Me_3SiNHNHSiMe_3 \xrightarrow[\text{MeI}]{\textit{n}\text{-BuLi}} Me_3SiN-NSiMe_3 + (Me_3Si)_2NNMe_2$$
$$\underset{\substack{| \quad | \\ Me \ Me}}{}$$

Ordinary hemiacetals are less stable than acetals and, in fact, they have the tendency to disproportionate to give the parent carbonyl compounds and the acetals. Hemithioacetals are even less stable in comparison with hemiacetals and dithioacetals.

The idea of symbiosis has been extended to S_N2 reactions by Pearson and Songstad (80). Analysis of rate data from the following reactions in methanol yielded a pattern consistent with symbiotic stabilization of the five-coordinated transition state. Later, it was demonstrated that the symbiotic effect is even larger in aprotic solvents such as acetonitrile (81). Gas phase S_N2 reactions also display this effect (82). Leaving group symbiosis is important in determining the alkylation site of acetoacetic esters (83). Apparently transition state symbiosis also operates in aromatic substitution ($S_N Ar$) (84).

$$Nu^\ominus + MeOTs \longrightarrow NuMe + TsO^\ominus$$

$$Nu^\ominus + MeI \longrightarrow NuMe + I^\ominus$$

$$Nu^\ominus = \text{nucleophile}$$

The symbiotic effect is not always the dominating one and other factors can take precedence over it. A case in point is the halogen exchange between CFI_3 and ClF_3 in the gaseous state. The process is endothermic by 50 kcal, presumably due to the considerably large steric strain of tetraiodomethane which has a very positive heat of formation (1).

A note of caution has been issued by Ahrland (85) who states that the co-ordination of very soft ligands to a [soft] (b)-metal ion acceptor will decrease its (b)-character, or even turn it into a [hard] (a)-acceptor. In the realm of organic chemistry this factor can be neglected.

3.3. INTRINSIC STRENGTH

Pearson (1) states in his treatment of the generalized acid–base reaction that formation of the complex depends on four parameters: the intrinsic strength S and the softness σ of both the acid and the base.

Although we usually concentrate on the influence of softness in discussions, the equally important strength factor should also be considered. Fortunately it seems that strength and softness work together in most cases, and the former factor is therefore often taken for granted.

Flagrant examples which violate the HSAB principle are known. Most of these can be attributed to the inherent strength of the acid–base complex which overrides the match in softness. That the extremely soft Lewis base $H^{\ominus}$ combines with the very hard acid $H^{\oplus}$ exothermically to give the stable hydrogen molecule is a case in point.

In the successive oxidation of methane, exothermicity increases with each step until the last. Pearson and Songstad (1) suggested that the acid $HOCO^{\oplus}$ is very strong, hence it would rather unite with the soft $H^{\ominus}$ which is stronger than the hard $OH^{\ominus}$. Consequently, formic acid is unusually stable against further oxidation. It should be noted that the conversion to carbonic acid is favored on the basis of symbiosis.

$$CH_4 \longrightarrow CH_3OH \longrightarrow CH_2(OH)_2 \longrightarrow HCOOH \longrightarrow H_2CO_3$$

Wittig reagents attack the harder sulfonyl center instead of F in sulfonyl fluorides (86). The strong polarized S–F bond forces the soft ylidic carbanion to interact with the harder acceptor.

$$RSO_2F + CH_2{=}PPh_3 \longrightarrow RSO_2CH_2\overset{\oplus}{P}Ph_3 \; F^{\ominus} \xrightarrow{CH_2=PPh_3} RSO_2CH{=}PPh_3 + Ph_3\overset{\oplus}{P}Me \; F^{\ominus}$$

A similar reaction takes place with dimethyloxosulfonium methylide (87). On the other hand, sulfonyl chlorides react with the Wittig reagent in accordance with the HSAB principle.

$$RSO_2Cl + CH_2{=}PPh_3 \longrightarrow RSO_2^{\ominus}ClCH_2\overset{\oplus}{P}Ph_3 \xrightarrow{CH_2=PPh_3} Ph_3P{=}CHCl$$
$$+ Ph_3\overset{\oplus}{P}Me \; PhSO_2^{\ominus}$$

REFERENCES

1. R. G. Pearson and J. Songstad, *J. Am. Chem. Soc.* **89**, 1827 (1967).
2. J. Rebek, Jr., S. F. Wolf, and A. B. Mossman, *Chem. Commun.* p. 711 (1974).
3. N. J. Bunce and N. G. Murray, *Tetrahedron* **27**, 5323 (1971).
4. E. Buncel and A. G. Davies, *J. Chem. Soc.* p. 1550 (1958).
5. B. Saville, *Angew. Chem., Int. Ed. Engl.* **6**, 928 (1967).
6. N. J. Leonard, *Rec. Chem. Prog.* **17**, 243 (1956).
7. N. J. Leonard, J. A. Adamcik, C. Djerassi, and O. Halpern, *J. Am. Chem. Soc.* **80**, 4858 (1958); N. J. Leonard, D. F. Morrow, and M. T. Rogers, *ibid.* **79**, 5476 (1957).
8. N. J. Leonard, T. W. Milligan, and T. L. Brown, *J. Am. Chem. Soc.* **82**, 4075 (1960).
9. N. J. Leonard and C. R. Johnson, *J. Am. Chem. Soc.* **84**, 3701 (1962).
10. N. J. Leonard and A. E. Yethon, *Tetrahedron Lett.* p. 4259 (1965); K. T. Go and I. C. Paul, *ibid.* p. 4265.
11. N. J. Leonard and W. L. Rippie, *J. Org. Chem.* **28**, 1957 (1963).
12. M. R. Bell and S. Archer, *J. Am. Chem. Soc.* **82**, 151 (1960).
13. J. Cantacuzene, R. Jantzen, and D. Ricard, *Tetrahedron* **28**, 717 (1972).
14. B. P. Susz, *Bull. Soc. Chim. Fr.* p. 2671 (1965).
15. A. T. Nielsen and W. J. Houlihan, *Org. React.* **16**, 1 (1968).
16. J. E. Dubois and P. Felimann, *Tetrahedron Lett.* p. 1225 (1975).
17. J. L. Burdett and M. T. Rogers, *J. Am. Chem. Soc.* **86**, 2105 (1964).
18. E. Allenstein and A. Schmidt, *Spectrochim. Acta* **20**, 1451 (1964).
19. E. Allenstein and P. Quis, *Chem. Ber.* **97**, 3163 (1964).
20. S. W. Peterson and J. M. Williams, *J. Am. Chem. Soc.* **88**, 2866 (1966).
21. K. Wiechert, H.-H. Heilmann, and P. Mohr, *Z. Chem.* **3**, 308 (1963).
22. K. M. Malony and B. S. Rabinovitch, *in* "Isonitrile Chemistry" (I. Ugi, ed.), Chapter 3. Academic Press, New York, 1971.
23. S. Yamada, K. Takashima, T. Sato, and S. Terashima, *Chem. Commun.* p. 811 (1969).
24. S. Yamada, M. Shibasaki, and S. Terashima, *Chem. Commun.* p. 1008 (1971).
25. A. Fava, A. Iliceto, A. Ceccon, and P. Koch, *J. Am. Chem. Soc.* **87**, 1045 (1965).
26. B. B. Wayland and R. H. Gold, *Inorg. Chem.* **5**, 154 (1966).
27. R. West, D. L. Powell, M. K. T. Lee, and L. S. Whatley, *J. Am. Chem. Soc.* **86**, 3227 (1964).
28. R. West, D. L. Powell, L. S. Whatley, M. K. T. Lee, and P. von R. Schleyer, *J. Am. Chem. Soc.* **84**, 3221 (1962).
29. M. H. Krackov, C. M. Lee, and H. G. Mautner, *J. Am. Chem. Soc.* **87**, 892 (1965).
30. L. D. Pettit, A. Royston, C. Sherrington, and R. J. Whewell, *Chem. Commun.* p. 1179 (1967).
31. W. A. G. Graham and F. G. A. Stone, *J. Inorg. Nucl. Chem.* **3**, 164 (1956).
32. J. P. Laurent, *Ann. Chim. (Paris)* [13] **6**, 677 (1961).
33. R. A. Braun, D. C. Brown, and R. M. Adams, *J. Am. Chem. Soc.* **93**, 2823 (1971).
34. D. Sheehan, Ph.D. Thesis, Yale University, New Haven, Connecticut (1964), quoted by A. J. Leusink, W. Drenth, J. G. Noltes, and G. J. M. van der Kerk, *Tetrahedron Lett.* p. 1263 (1967).
35. D. E. Young, G. E. McAchran, and S. G. Shore, *J. Am. Chem. Soc.* **88**, 4390 (1966).
36. A. H. Cowley, *Chem. Rev.* **65**, 617 (1965).
37. A. B. Burg and P. J. Slota, *J. Am. Chem. Soc.* **80**, 1107 (1958).
38. N. L. Smith and H. H. Sisler, *J. Org. Chem.* **28**, 272 (1963).
39. G. A. Ozin, *Chem. Commun.* p. 104 (1969).
40. J. Dahlmann, A. Rieche, and L. Austenat, *Angew. Chem., Int. Ed. Engl.* **5**, 727 (1966).
41. S. H. Smallcombe and M. C. Caserio, *J. Am. Chem. Soc.* **93**, 5826 (1971).
42. P. Nicpon and D. W. Meek, *Chem. Commun.* p. 398 (1966).

43. B. C. Mayo, *Chem. Soc. Rev.* **2**, 49 (1973).
44. A. F. Cockerill, G. L. O. Davies, R. C. Harden, and D. M. Rackham, *Chem. Rev.* **73**, 553 (1973).
45. D. H. Williams, *Pure Appl. Chem.* **40**, 25 (1974).
46. A. K. Bose and P. R. Srinivasan, *Tetrahedron Lett.* p. 1571 (1975).
47. K. Flatau and H. Musso, *Angew. Chem., Int. Ed. Engl.* **9**, 379 (1970).
48. J. F. Wolf, P. G. Harch, R. W. Taft, and W. J. Hehre, *J. Am. Chem. Soc.* **97**, 2902 (1975).
49. J. Jerkunica and T. G. Traylor, *J. Am. Chem. Soc.* **93**, 6278 (1971).
50. G. A. Olah and P. R. Clifford, *J. Am. Chem. Soc.* **93**, 1261 and 2320 (1971).
51. M. L. Ziegler, *Angew. Chem., Int. Ed. Engl.* **7**, 222 (1968).
52. G. W. Astrologes and J. C. Martin, *J. Am. Chem. Soc.* **97**, 6909 (1975).
53. D. B. Denney, D. Z. Denney, and Y. F. Hsu, *J. Am. Chem. Soc.* **95**, 4064 (1973).
54. E. O. Fischer and A. Maasbol, *Chem. Ber.* **100**, 2445 (1967).
55. E. O. Fischer and A. Riedel, *Chem. Ber.* **101**, 156 (1968).
56. P. E. Baikie, E. O. Fischer, and O. S. Mills, *Chem. Commun.* p. 1199 (1967).
57. E. O. Fischer and H. J. Kollmeier, *Angew. Chem., Int. Ed. Engl.* **9**, 309 (1970).
58. M. L. H. Green and C. R. Hurley, *J. Organomet. Chem.* **10**, 188 (1967).
59. E. O. Fischer, G. Kreis, C. G. Kreiter, J. Müller, G. Huttner, and H. Lorenz, *Angew. Chem., Int. Ed. Engl.* **12**, 564 (1973).
60. P. Yates, *J. Am. Chem. Soc.* **74**, 5376 (1952).
61. E. Müller, B. Zech, and H. Kessler, *Fortsch. Chem. Forsch.* **7**, 128 (1966).
62. D. S. Crumrine, T. J. Haberkamp, and D. J. Suther, *J. Org. Chem.* **40**, 2274 (1975).
63. R. G. Pearson, H. B. Gray, and F. Basolo, *J. Am. Chem. Soc.* **82**, 787 (1960).
64. C. Rüchardt and G. N. Schrauzer, *Chem. Ber.* **93**, 1840 (1960).
65. T.-L. Ho, M. Henninger, and G. A. Olah, *Synthesis* p. 815 (1976).
65a. L. A. P. Kane-Maguire, F. Basolo, and R. G. Pearson, *J. Am. Chem. Soc.* **91**, 4609 (1969).
66. A. W. Johnson, "Ylid Chemistry." Academic Press, New York, 1966.
67. L. A. Kaplan and H. B. Pickard, *Chem. Commun.* p. 1500 (1969).
68. B. L. Dyatkin, S. R. Sterlin, B. I. Martynov, and I. L. Knunyants, *Tetrahedron Lett.* p. 345 (1971).
69. J. Hine and A. M. Dowell, *J. Am. Chem. Soc.* **76**, 2688 (1954).
70. J. Hine and D. C. Duffey, *J. Am. Chem. Soc.* **81**, 1131 (1959).
71. H. Sakurai, A. Hosomi, and M. Kumada, *Chem. Commun.* p. 521 (1969).
72. C. K. Jørgensen, *Inorg. Chem.* **3**, 1201 (1964).
73. J. Hine, *J. Am. Chem. Soc.* **85**, 3239 (1963).
74. L. O. Brockway, *J. Phys. Chem.* **41**, 185 and 747 (1937).
75. J. Hine, *J. Am. Chem. Soc.* **85**, 3239 (1963).
76. F. Terrier, G. Ah-Kow, M.-J. Pouet, and M.-P. Simmonin, *Tetrahedron Lett.* p. 227 (1976).
77. A. Finch, P. J. Gardner, A. Hameed, and K. K. SenGupta, *Chem. Commun.* p. 854 (1969).
78. L. L. Chang and D. B. Denney, *Chem. Commun.* p. 84 (1974).
79. R. E. Bailey and R. West, *J. Am. Chem. Soc.* **86**, 5369 (1964).
80. R. G. Pearson and J. Songstad, *J. Org. Chem.* **32**, 2899 (1967).
81. L. B. Engemyr and J. Songstad, *Acta Chem. Scand.* **26**, 4179 (1972).
82. J. I. Brauman, W. N. Olmstead, and C. A. Lieder, *J. Am. Chem. Soc.* **96**, 4030 (1974).
83. A. L. Kurts, N. K. Genkina, A. Macias, I. P. Beletskaya, and O. A. Reutov, *Tetrahedron* **27**, 4777 (1971).
84. D. E. Giles and A. J. Parker, *Aust. J. Chem.* **26**, 273 (1973).
85. S. Ahrland, *Struct. Bonding (Berlin)* **1**, 207 (1966).
86. A. M. van Leusen, B. A. Reith, A. J. W. Iedema, and J. Strating, *Recl. Trav. Chim. Pays-Bas* **91**, 37 (1972).
87. W. E. Truce and G. D. Madding, *Tetrahedron Lett.* p. 3681 (1966).

4

Displacement Reactions

4.1. GENERAL MECHANISMS

Of the various kinds of organic reactions substitution, especially aliphatic nucleophilic substitution, is most amenable to HSAB correlation. The transition state of an S_N process may be envisaged as an acid–base pairing, the nucleophile being equated to a donor base and the electrophilic center to an acceptor acid.

Mechanistically there are two extremes, namely, the S_N1 and the S_N2 types of reaction according to the Ingold–Hughes designation (1). Winstein *et al.* (2) classified the reactions as *N* and *Lim*; they do not recognize S_N1 and S_N2 reactions as distinct processes.

The average relative rates for S_N2 reactions (3) are Me, 30; Et, 1; *n*-Pr, 0.4; *n*-Bu, 0.4; *i*-Pr, 0.025; *i*-Bu, 0.03; neopentyl, 10^{-5}; allyl, 40; and benzyl, 120.

The escalated reactivity of allylic and benzylic compounds originates from an $-I$ effect. Heterosubstituents in the α position tend to enhance reaction rates through the assistance of ionization. Substrates having $-I$ and $-M$ groups such as cyano and carbonyl in the α position resist ionization. However, these compounds are very reactive in S_N2 processes. In terms of the HSAB concept, the heightened reactivity is simply due to the hardening of the reaction centers. It will be seen that the rate increase in the S_N2 reactions for the α-carbonyl substrates pertains only to those involving hard nucleophiles.

$$ROCH_2X \longrightarrow ROCH_2^{\oplus} \; X^{\ominus} \longleftrightarrow R\overset{\oplus}{O}{=}CH_2 \; X^{\ominus}$$

In a theoretical investigation of the S_N2 reactions three rules were formulated by Harris and Kurz (4). One of these rules states that if an S_N2 transition state involves nucleophilic atoms from different rows of the Periodic Table an electron-withdrawing substituent at the central atom will tend to increase the order of the reacting bond to the lighter atom, and decrease the order of the other reacting bond. In other words, if the central atom is harder, it will bind more tightly to the harder nucleophile.

The S_N2 opening of strained heterocycles provides insights into many reactivity parameters. Treatment of fluoroepoxides with hydrogen fluoride leads exclusively to *gem*-difluorides (5). The simplest explanation for this regiospecificity appears to be the symbiotic effect.

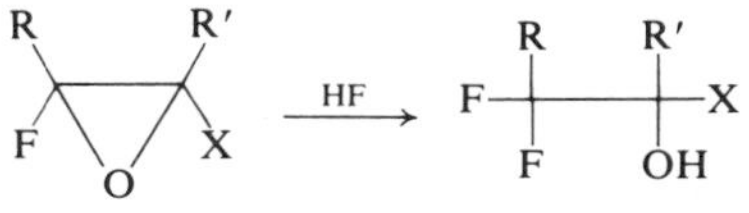

$$X = H, CN$$

Whereas epoxides are readily cleaved by ammonia and amines (6), ring opening of thiiranes is rather sluggish. The situation is ameliorated by adding silver ion (7) to coordinate with the sulfur atom which then renders the ring C–S bond more permanently polarized (and hence *less polarizable*), i.e., the ring carbon becomes harder and more responsive to hard bases. Precedence for this activation is found in a one-step synthesis of sulfenamides from disulfides (8).

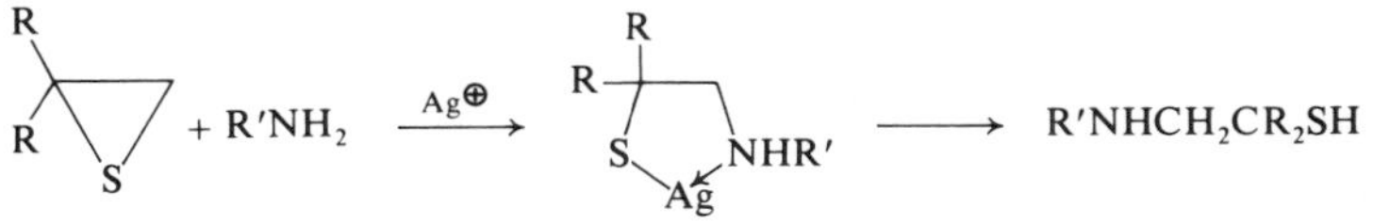

$$RSSR + 2R_2'NH + AgX \longrightarrow RSNR_2' + RSAg + R_2'N^\oplus H_2 \ X^\ominus$$

4.2. NUCLEOPHILIC REACTIVITY

A nucleophile is a reagent which supplies an electron pair to form a new bond between itself and another atom (9). Swain and Scott (10) proposed that in describing properties of these species "basicity" be used solely in equilibria (thermodynamic) phenomena and "nucleophilicity" in rate (kinetic) phenomena. Parker (11) further suggested that the thermodynamic affinity for elements other than hydrogen be termed M-basicity, e.g., carbon basicity or sulfur basicity.

By comparing the rate constant k of a particular reaction in a series with a reference reaction whose rate constant is k_0, Swain and Scott established (10) the nucleophilicity (n) of the reagent as

$$\log(k/k_0) = sn$$

where s is a measure of substrate susceptibility to the nucleophile. It is defined as 1.0 for methyl bromide at $25°$. The value of n for water is set at zero.

This correlation does not hold when different types of reactions are analyzed.

Edwards (12) has formulated a four-parameter equation comprising a nucleophilicity term and a basicity term for correlating electron donors with rates and equilibria.

$$\log(k/k_0) = \alpha E_n + \beta H$$

Nucleophilicity has also been correlated with basicity and polarizability (13) by

$$\log(k/k_0) = AP + BH$$

The relationship of E_n to the nucleophilic constant n of the Swain–Scott equation appears to be linear.

Factors influencing nucleophilicity have been discussed and summarized (14, 15). For $S_N2(C)$ reactions Edwards and Pearson (14) have compiled a nucleophilic order: $RS^\ominus > ArS^\ominus > S_2O_3^{2\ominus} > (H_2N)_2CS > I^\ominus > CN^\ominus > SCN^\ominus > OH^\ominus > N_3^\ominus > Br^\ominus > ArO^\ominus > Cl^\ominus > C_5H_5N > AcO^\ominus > H_2O$. They noted that polarizability is the dominant factor. Since polarizability is intrinsically associated with chemical softness of a species, parallelism exists between nucleophilic order and softness order.

During a study of the displacement of o-substituted benzyl chlorides, Bunnett and Reinheimer (16) discovered that the reactivity of $I^\ominus$ and $C_6H_5S^\ominus$, on the one hand, and $MeO^\ominus$, on the other, depends on the polarizability of the ortho substituent.

The rate ratio $k_{PhS^\ominus}/k_{MeO^\ominus}$ for substitution varies from 512 for p-methoxy to 5300 for p-nitrobenzyl bromides (17). The former compound exhibits considerable carbenium ion character in the transition state, hence it reacts faster with $MeO^\ominus$ (hard–hard interaction).

The true nucleophilic order for halide ions was at one time nebulous. For example, in the displacement of n-butyl brosylate, two series of halides display opposite reactivity: $Bu_4NCl > Bu_4NBr > Bu_4NI$; but $LiI > LiBr > LiCl$. Winstein *et al.* (18) ascribed the noncorrespondence of lithium salt reactivity to the intrinsic strength and increasing dissociation of salts with larger anions.

The apparent intrinsic nucleophilicities of halide ions have been determined by the thermal decomposition of quaternary ammonium salts (19). Complications due to solvation are eliminated. It should be noted that the leaving group is a hard base.

$$(n\text{-}C_5H_{11})_4N^{\oplus} \ X^{\ominus} \ \xrightarrow{\ \Delta\ } \ (n\text{-}C_5H_{11})_3N + n\text{-}C_5H_{11}X$$

$$k_X^{rel} \ \text{for } Cl:Br:I = 620:7.7:1$$

In the two-phase alkylation using alkali hydroxides as bases and quaternary ammonium halides as catalysts (20) a reverse order (RCl > RBr > RI) of the customary alkylator reactivity prevails. Quaternary ammonium iodides exhibit little catalytic effect (21). Because the effective concentration of $R_4N^{\oplus}Nu^{\ominus}$ is low and limited by the initial amount of catalyst, the nucleophile cannot be expected to compete successfully for the alkylating agent with $I^{\ominus}$, if its softness lies in between $OH^{\ominus}$ and $I^{\ominus}$.

One of the strongest nucleophiles known is the complex-bound Co(I) ion (22, 23). In this form vitamin B_{12} mediates a myriad of remarkable transformations. Methylation of Ni(II) corrole anions occurs at the nickel (24).

A completely different picture emerges when nucleophilic attack on the carbonyl carbon is examined. The nucleophilic order follows the basicity order more closely as demonstrated by reactions of *p*-nitrophenyl acetate with anilines, pyridines, imidazoles, oxyanions, etc. (25, 26). The same is true for reactions with ethyl chloroformate (27).

The attack on the carbonyl proceeds via an addition–elimination mechanism in which the nucleophile acquires a partial positive charge prior to the transition state. The less encumbered electrophilic site and the need to involve only the substrate *p* orbitals permit closer approach of the attacking species. This situation resembles the close range capture of a small proton by a Brønsted base.

Because the transition state structure is tight, the steric effect may become an important factor influencing nucleophilicity. The reduced steric demands around the nitrogen atoms of aziridines and azetidines are responsible, at least in part, for the greatly enhanced nucleophilicity toward ester carbonyl as compared with other amines having similar basicities (28).

4.3. FORMATION AND CLEAVAGE OF *O*-ALKYL BOND OF CARBOXYLIC ESTERS

The esterification of carboxylic acids with diazoalkanes has become a classical method. The success of the reaction stems from the interaction of two rather hard species which are simultaneously generated by proton transfer from the carboxylic acid to the α carbon of the diazoalkanes.

$$RCOOH + R'CHN_2 \rightarrow RCOO^\ominus \ R'CH_2N_2^\oplus \rightarrow RCOOCH_2R' + N_2$$

The combination of alkali metal salts of carboxylic acids and alkyl halides seldom finds use in ester synthesis owing to low yields. The crucial O–C bond formation is a hard–soft interaction. In contrast, the efficient reaction of acyl halides with alcohols consists of a hard–hard interaction.

Improved techniques for ester formation invariably involve better soft–hard compatibility of the reaction partners, e.g., replacement of alkyl halides with the harder dialkyl sulfates (29) or softening the carboxylate ion to accommodate the alkyl halides. The carboxylates of Cu(I), (30–32), Tl(I), Ag(I) (33), and Hg(II) (34) have been successfully converted into esters.

Solvents also play an important role in regulating the reactivity of anions. Hexamethylphosphoric triamide (HMPT) solvates alkali metal ions so strongly that the counteranions are practically free. Under such conditions the inherent bond strengths determine the ease of reaction. Since the O–C bond is stronger than the C–X bonds (X = Cl, Br, I), facile displacement of halide ions by carboxylates in HMPT is a logical consequence (35–38).

Hindered esters which are difficult to hydrolyze by the standard methods may be cleaved utilizing soft nucleophiles which attack the alkoxy carbon atoms specifically. Alkane thiolates are very efficient agents for the hydrolysis of phenacyl (39), 9-anthrylmethyl (40), and methyl esters (41). The inertness of higher esters is largely due to steric effects, as it has been demonstrated that ω-haloalkyl esters can be cleaved by an anchoring method (42–44).

Thiolate ions undergo S-alkylation on reaction with trichloroacetic esters in contrast to normal solvolysis (45) of trifluoro- and trichloroacetates by hard bases. In refluxing dimethylformamide (DMF) the thiocyanate ion cleaves methyl and benzyl esters (46). Alternatively, a eutectic melt of NaSCN–KSCN may be employed (47). The neutral mixture from the latter reaction consists of 96% MeSCN and 4% MeNCS, indicating a predominant displacement by the softer S terminus of the base.

It is well known that $I^\ominus$ is one of the softest common nucleophiles. The use of $I^\ominus$ in deblocking methyl esters (Taschner–Eschenmoser method) (48–51) has become a conventional procedure in organic synthesis. Nucleophilic attack by $I^\ominus$ on the methyl group $[S_N 2(C)]$ is undisputedly operative.

4.4. MULTICENTERED REACTIONS: SAVILLE'S RULES

During an analysis of bimolecular reactions Saville (52) stated that the HSAB principle might be applied to most organic processes. As substitution is assisted by precoordination of the leaving group with an electrophile, it becomes evident that a general form of the four-centered reaction involving cooperative action of a nucleophile (Na) and an electrophile (E) on a substrate (push-pull mechanism) should proceed with greater facility. Addition to π-electron systems works in the same way.

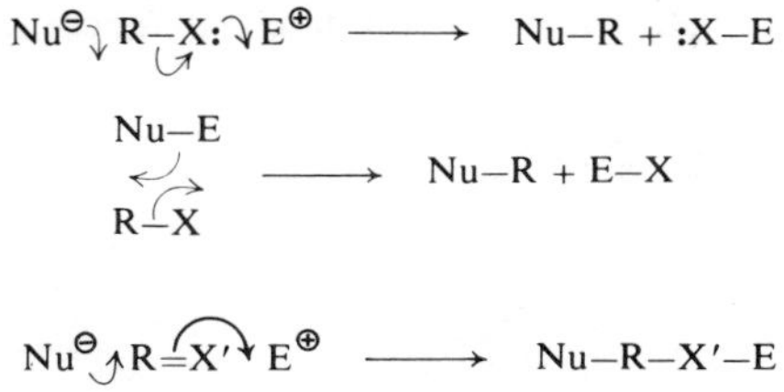

$$Nu^{\ominus} \; R{-}X: \; E^{\oplus} \longrightarrow Nu{-}R + :X{-}E$$

$$\begin{array}{c} Nu{-}E \\ R{-}X \end{array} \longrightarrow Nu{-}R + E{-}X$$

$$Nu^{\ominus} \; R{=}X' \; E^{\oplus} \longrightarrow Nu{-}R{-}X'{-}E$$

Saville deduced that the most effective way to achieve the desirable bond cleavage would be to provide a nucleophilic and an electrophilic species of the same (hard or soft) category as the acid A and the base B, respectively, as shown below.

Rule 1:

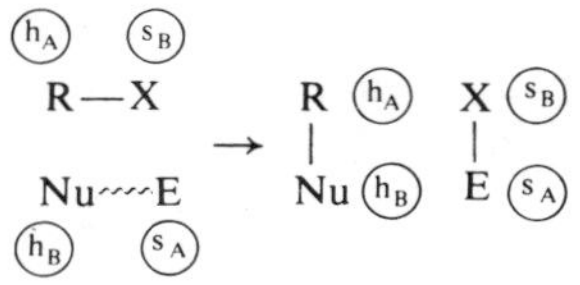

$$\begin{array}{cc} \textcircled{h}_A & \textcircled{s}_B \\ R{-}X \\ Nu{\sim}E \\ \textcircled{h}_B & \textcircled{s}_A \end{array} \rightarrow \begin{array}{cc} R\,\textcircled{h}_A & X\,\textcircled{s}_B \\ | & | \\ Nu\,\textcircled{h}_B & E\,\textcircled{s}_A \end{array}$$

Rule 2:

$$\begin{array}{cc} \textcircled{s}_A & \textcircled{h}_B \\ R{-}X \\ Nu{\sim}E \\ \textcircled{h}_B & \textcircled{s}_A \end{array} \rightarrow \begin{array}{cc} R\,\textcircled{s}_A & X\,\textcircled{h}_B \\ | & | \\ Nu\,\textcircled{s}_B & E\,\textcircled{h}_A \end{array}$$

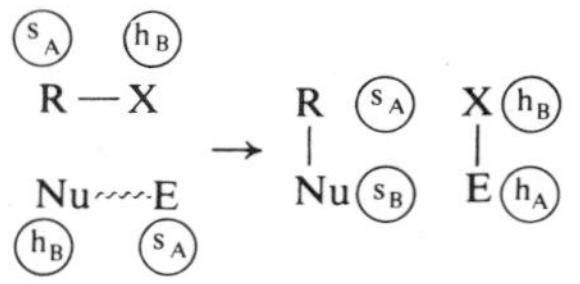

Optimal conditions will be achieved when the substrate bond R$-$X possesses maximal hard$-$soft dissymmetry.

In his original paper (52) Saville enumerated a great number of cases to substantiate his viewpoint. His discussion of rule 1 extends to C$-$B bond cleavage, reactions of aldehydes, thioacids and esters, phosphoryl and sulfonyl compounds. For rule 2 he gives examples of C$-$O bond fission, substitution at oxygen and sulfur, etc. Some of these are reiterated here.

A most striking phenomenon is the heterolysis of molecular hydrogen according to the following equation.

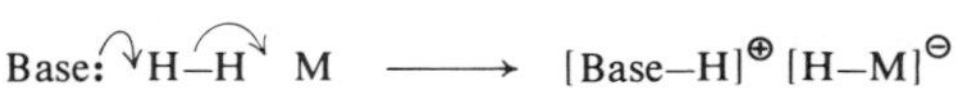

$$Base: \; H{-}H \; M \longrightarrow [Base{-}H]^{\oplus} \, [H{-}M]^{\ominus}$$

Only the combination of a hard base (e.g., carboxylate ion, pyridines) and a soft metal cation (e.g., Ag^I, Hg^I, Hg^{II}, Cu^I, Pt^{II}) is suitable (53, 54). Saville's rule 1 is faithfully obeyed.

Methoxycarbonylmercury compounds undergo acid-catalyzed decomposition in the presence of "assistors" of a soft nature, e.g., RSH, $I^\ominus$, and $Br^\ominus$ (55). In the von Braun degradation (56) tertiary amines are ruptured on exposure to cyanogen bromide to yield bromocyanamides, in accordance with Saville's rule.

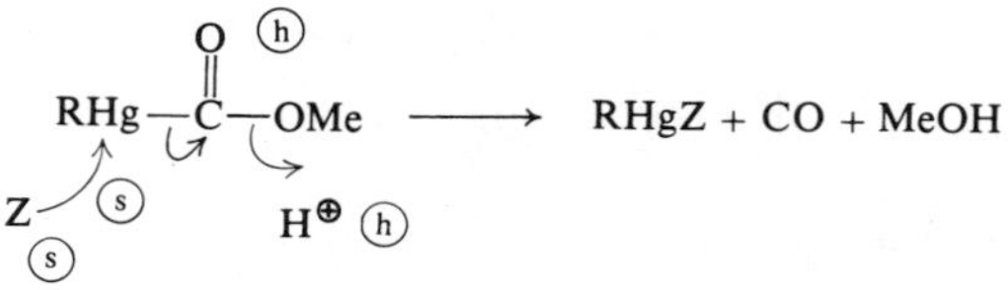

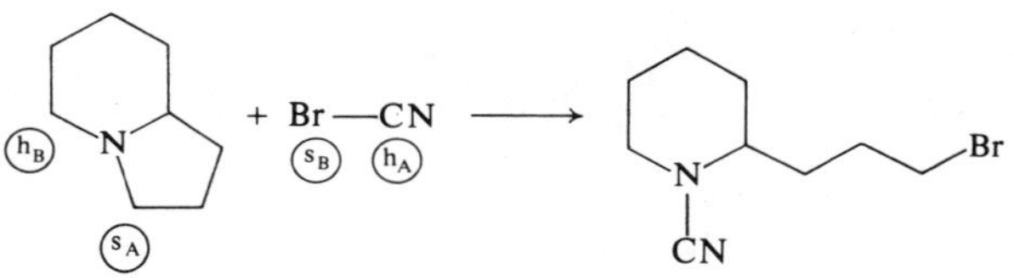

Oxidation of the iodide ion with acidic H_2O_2 (57, 58) is shown in the following equations.

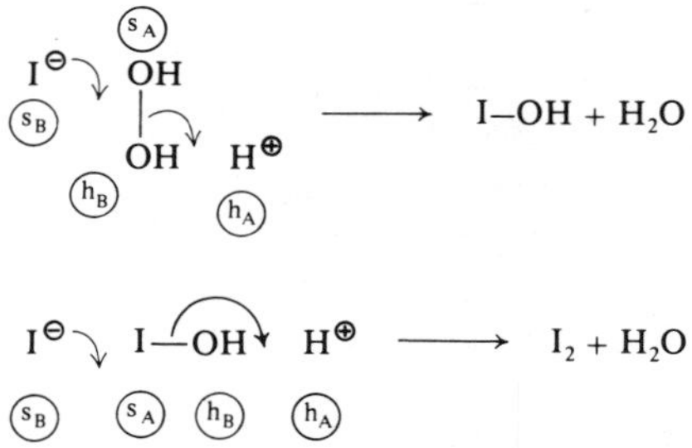

The classical Zeisel method for the determination of ether (59) consists of digesting the substrate with hydrogen iodide. Analysis shows that the reaction proceeds in a favorable manner as designated by Saville. Sulfides do not interfere with the determination because the hard–soft combination is incorrect in such cases.

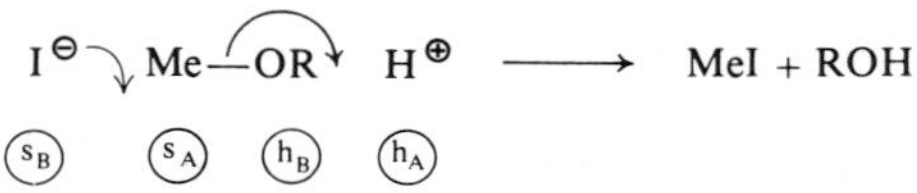

The simplest procedure for the conversion of alkanols to alkyl halides is by treatment with hydrogen halides. That alkyl fluorides cannot be prepared in a similar manner is attributable to the hardness of the anion.

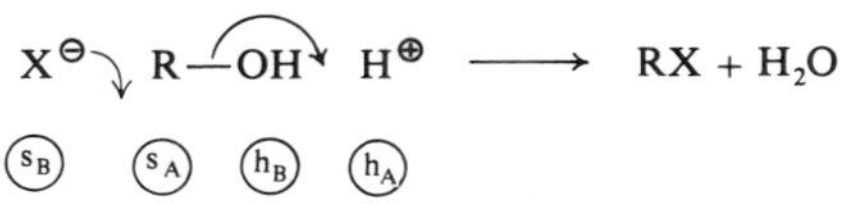

The same principle is involved in the preparation of certain phosphonium salts (60) and S-alkyl thiouronium salts (61) by reacting alcohols with phosphines and thiourea, respectively, under acid catalysis.

Saville's rules serve to indicate optimal catalytic conditions. However, there exist many facile multicenter reactions which do not fall into either of the two categories. The hydrolysis rates for acetylimidazole (62) rise with increasing imidazole buffer concentration, and this observation suggests a ⓗ...ⓗ–ⓗ...ⓗ transition state.

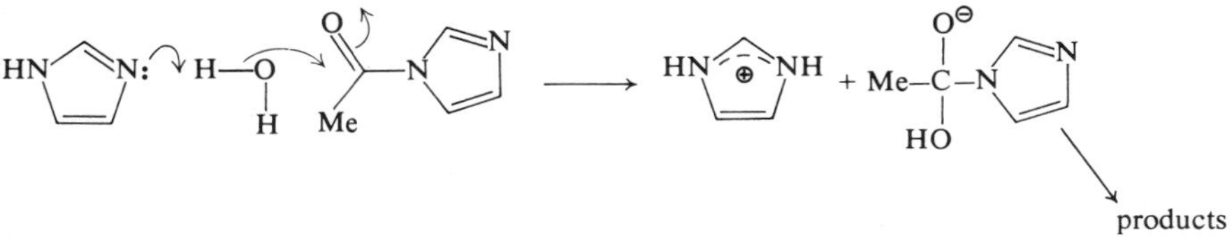

products

Hydrolysis of methyl methylene phosphate (63) has been similarly interpreted.

It has been shown that syn-p-nitrophenyl phenacyl methylphosphonate oxime undergoes rapid hydrolysis (64) in which the oximate ion acts as an internal base.

The addition of N-methylaniline to aryl isocyanate is catalyzed by phenol and the addition of phenol to the isocyanates is likewise catalyzed by N-methylaniline (65). Cyclic transition states have been proposed for these reactions. All these interactions are hard.

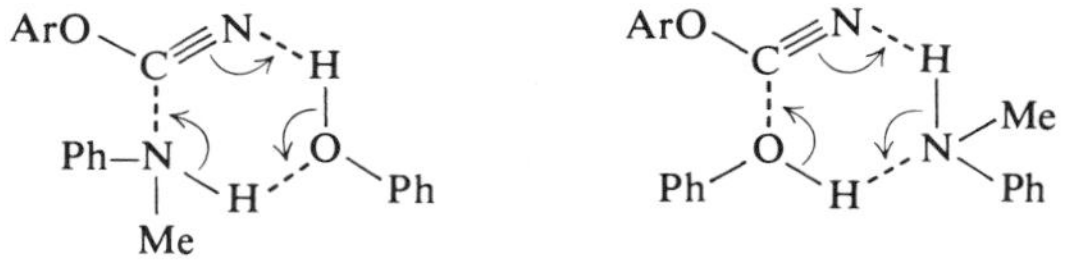

2-Hydroxypyridine is a very useful catalyst for promoting aminolysis of esters (66). The process involves perturbation at eight centers with all the bond-forming termini interacting with one another by hard forces. The detailed atomic arrangement in the transition state of 2-hydroxypyridine-catalyzed mutarotation of α-D-tetramethylglucose (67, 68) in benzene is not clear; a similar type of interaction undoubtedly takes place, however.

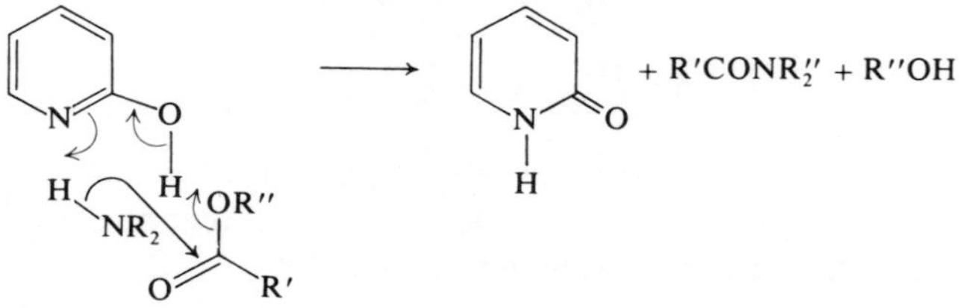

Although Saville rule 1 is shown in the first and second steps of the reaction of silver benzoate with carbon disulfide (69), the decomposition of the intermediary O,O-dibenzoylthiocarbonate proceeds with (h) ··· (h)–(h) ··· (h) reorganization.

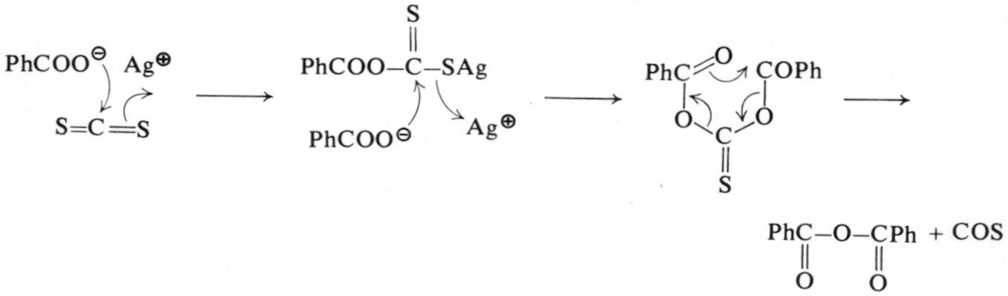

Four-centered reactions in which all the acid–base pairings are made up of soft species are also known. Some examples are given below.

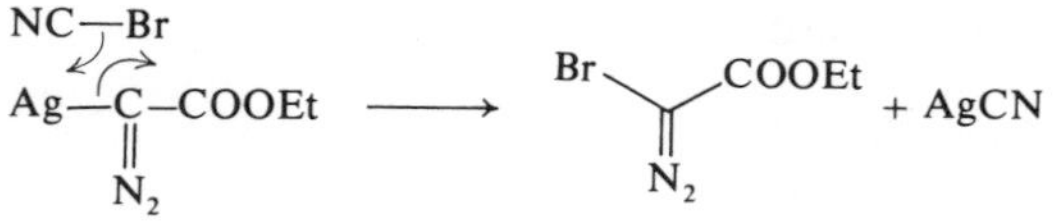

(Ref. 70)

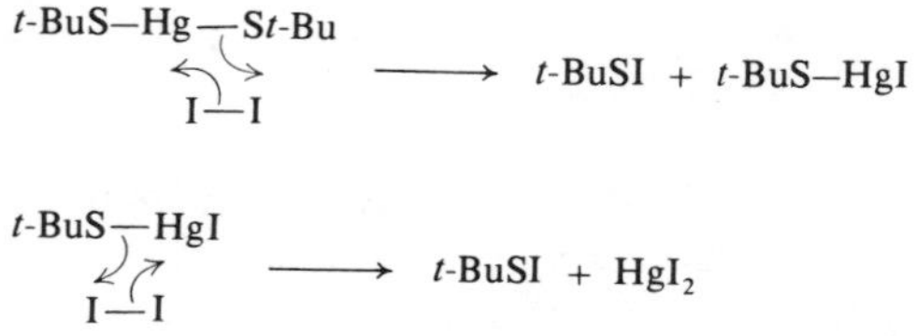

(Ref. 71)

4.5. AMBIDENT REACTIVITY

Chemical species may possess two or more reactive sites which are often interrelated by tautomerism or mesomerism. Such entities are called *ambident* (72). They differ fundamentally from simple bifunctional compounds.

4.5.1. Ambident Electrophiles

Many ambident electrophiles are energetic species whose reactions have been reviewed by Hünig (73). Their reaction selectivities closely follow the HSAB principle e.g.,

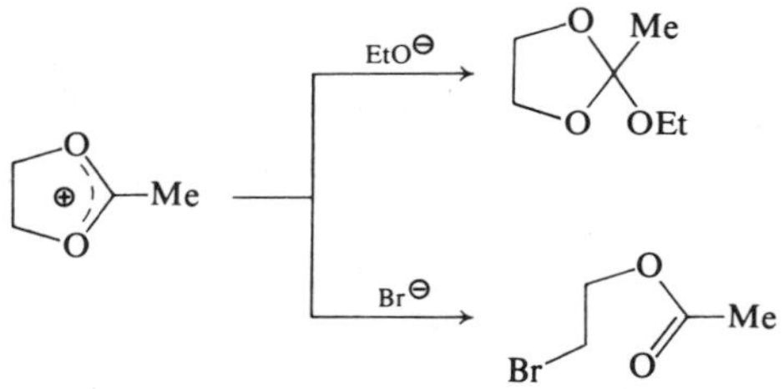

Dimethylformamide dimethyl acetal undergoes alkoxy exchange when dissolved in alcohols. On the other hand, it is "hydrolyzed" on treatment with alkanethiols (74).

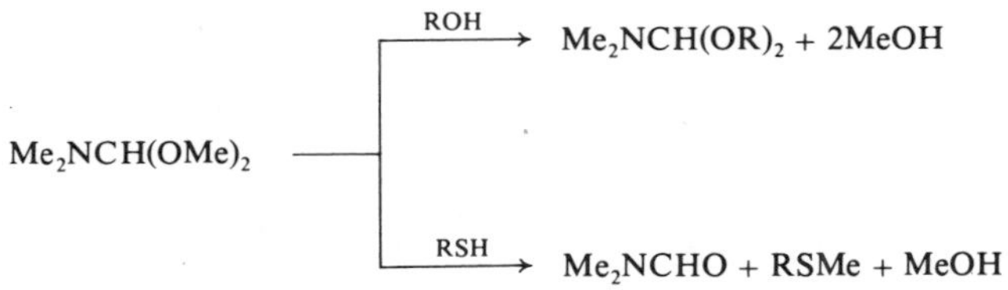

The behavior of tris(methylthio)methyl cation $(MeS)_3C^\oplus$ toward various nucleophiles (75) can be explained by the HSAB principle. Water neutralizes the carbenium center, soft ions ($Br^\ominus$, $I^\ominus$, $CN^\ominus$) attack the methyl group, and methanethiol also adds to the central carbon atom as dictated by the symbiotic effect. Triphenylphosphine reacts at the S site.

The somewhat similar 2-dialkylamino-1,3-dithiolinium salts react with hard bases ($OH^\ominus$, $ArNH_2$) at the carbenium site, with soft donors (e.g., $RS^\ominus$) at methylene carbon atoms resulting in ring opening (76).

Typical soft bases such as Me_2S (77) and $I^\ominus$ (78) attack the softer carbon atom of the *O*-alkyl group of alkoxydiazenium ions.

$$\overset{\oplus}{R_2N-N=OR'} \quad \longleftrightarrow \quad R_2\overset{\oplus}{N}=N-OR' \quad \longleftrightarrow \quad R_2N-\overset{\oplus}{N}-OR'$$

Heterocyclic quaternary azo salts such as the pyridinium derivatives offer many sites for nucleophilic reaction (79). Again the mode of reaction is determined by the softness of the attacking species.

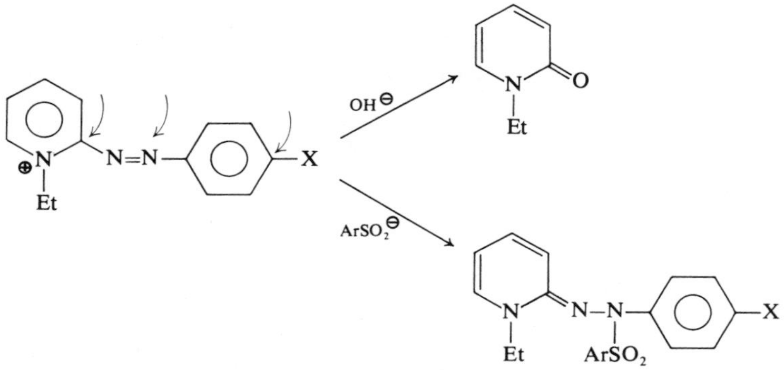

3,3-Dimethyl-1,2-dioxetane is decomposed by nucleophiles. Kinetic and product analyses indicate that azide ion displaces the peroxy group from carbon (80), whereas the bromide ion directly attacks the soft oxygen (81).

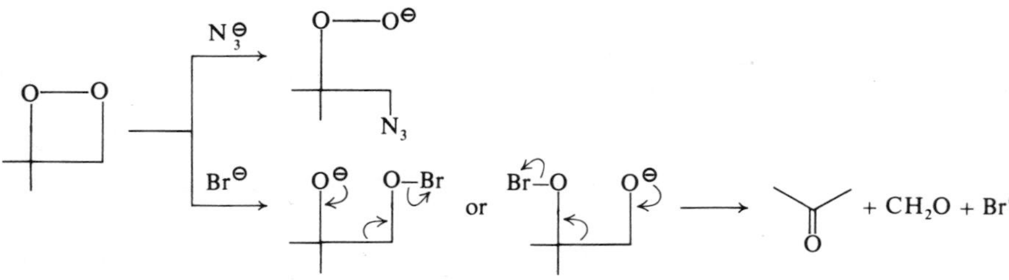

The highly reactive β-propiolactone displays discriminatory reaction modes toward bases of different softness. For example, the hard alkoxide ion opens the lactone ring via addition to the acyl carbon (82) whereas the soft cyanide (83) and thiolate ions (84) effect displacement at the β carbon. Indole is alkylated at its β position to give 3-(β-indolyl)propionic acid (85).

The cleavage of β-propiolactone with heterostannanes (86) apparently proceeds via complexes in which the ethereal oxygen of the lactone is (loosely) linked to the tin atom. A hard hetero group (e.g., OR, NR_2) is directed to bond with the acyl carbon. On the other hand, soft groups (SR, Hal) effect the formation of tin carboxylates.

The softer silicon and germanium atoms of diethylaminotrimethylsilane and -germane, respectively, impart a softer character to the nitrogen; thus, the interactions of these substances with β-propiolactone lead to β-dimethyl-aminopropionic acid derivatives. The diethylphosphino analogs Me_3M-PEt_2 (M = Sn, Si, Ge), in which the phosphorus atom is soft, promote C_β-O bond rupture of β-propiolactone exclusively (87).

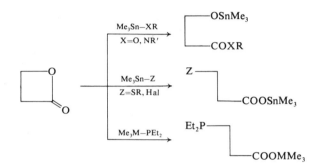

The soft base sodium benzenethiolate attacks p-nitrophenyl acetate only at the acyl carbon. On the other hand, its reaction with 2,4,6-trinitrophenyl benzoate follows an S_NAr pathway (88). The strong electron attraction of the trinitrophenoxy moiety renders the ester carbonyl much harder and more reluctant to interact with soft bases than usual. At the same time the aromatic ring becomes a good electron sink especially for soft bases.

N-Carbethoxyaziridine is an ambident electrophile with soft ring carbon sites and a hard carbonyl group. The HSAB principle may be applied to account for its reaction with a variety of nucleophilic agents (89).

$$\begin{array}{c} O \\ \parallel \\ \triangleright N-C-OEt \end{array}$$

(s) attack: $H^\ominus(NaBH_4)$, $PhNH_2$, Ph_3CLi
(h) attack: $H^\ominus(LiAlH_4)$, PNHLi, RLi
(s) + (h) attacks: Ph_2CHLi (soft 2 + hard 1)

Nitrosamines are ambident acceptors which can react at O, N, or α-H. The observation (90) that α-lithiation of Me_2NNO is most effective with the harder organolithium reagents (i-Pr$_2$NLi, 95%; PhLi, 80%; MeLi, 60%; n-BuLi, 45%; t-BuLi, 10%) agrees well with the HSAB expectation as the proton is harder than the oxygen and nitrogen electrophilic sites.

$$Me_2N-N=O + RLi \longrightarrow \overset{Me}{\underset{LiCH_2}{\Large >}}N-N=O + RH$$

Nitrobenzene and nitrosobenzene are attacked by aryllithium and Grignard reagents according to the following equations (91).

$$PhN=O \begin{cases} \xrightarrow{PhLi} \underset{Li^\oplus}{Ph\overset{\ominus}{N}-OPh} \xrightarrow[-PhOLi]{} Ph\ddot{N}: \xrightarrow{PhLi} Ph_2NLi \\ \\ \xrightarrow{PhMgX} Ph_2N-OMgX \end{cases}$$

$$PhNO_2 + ArLi \longrightarrow \underset{OLi}{PhN-OAr} \xrightarrow{ArLi} \underset{Ar}{PhN-OLi} + ArOLi$$

As acceptors both N and O of the nitroso group are quite soft.

In perchloryl fluoride (FClO$_3$) the fluorine is a soft acid. Formation of the C—F bond is observed when it is treated with carbanions. Thiolates are oxidized to disulfides by this reagent (92, 93). This latter reaction presumably involves sulfenyl fluoride intermediates. In contrast, alkoxide ions effect displacement at chlorine to afford perchlorate esters or further transformation products.

Since the reaction rates of phenyl chlorosulfate with various anions follow the order of $S_2O_3^{2\ominus} > CN^\ominus > I^\ominus > SO_3^{2\ominus} > SCN^\ominus > Br^\ominus > Cl^\ominus$, $F^\ominus$, $AcO^\ominus$, it is likely that displacement occurs at the soft chlorine atom (94).

N-Chlorosulfonylazetidinones possess three different electrophilic sites (95, 96). Soft bases (e.g., $I^\ominus$) would pull out the Cl, hard bases (RO$^\ominus$, R$_2$NH) tend to attack the lactam carbonyl and the sulfonyl S, as both are hard centers. The reaction with borderline azide ion gives rise to a mixture of a diazide from a hard-type attack and a cyclic urea which must arise from Cl abstraction, desulfonylation of the chlorosulfonylamide group, and simultaneous addition of a second azide ion to the carbonyl which triggers a Schmidt rearrangement. It seems that the fate (ring opening vs. rearrangement) of the tetrahedral azidohydrin intermediates depends on the nature of the N-substituent. If the substituent is a sulfinate anion or a negative charge, the ring-opening process becomes electronically unfavorable.

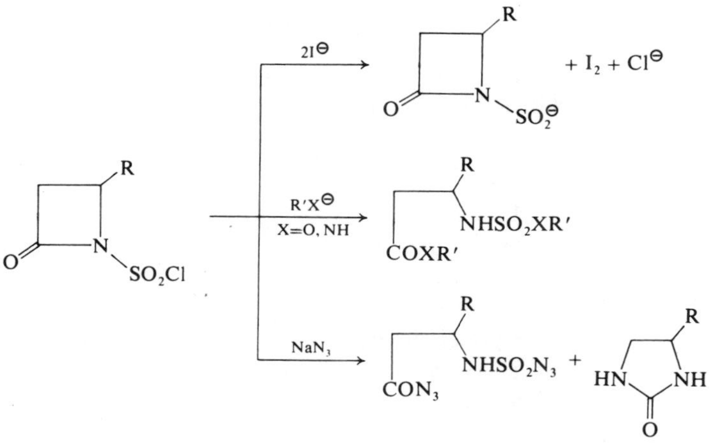

$2I^{\ominus}$ → [structure: R-substituted β-lactam with N–SO$_2^{\ominus}$] $+ I_2 + Cl^{\ominus}$

[structure: β-lactam with N–SO$_2$Cl]

$\dfrac{R'X^{\ominus}}{X=O,\,NH}$ → [structure] $NHSO_2XR'$, $COXR'$, R

NaN_3 → [structure] $NHSO_2N_3$, $\overset{|}{CON_3}$, R $+$ [imidazolidinone: HN–C(=O)–NH with R]

Another pertinent observation is the formation of diphenyl disulfide and N,N'-di(phenylthio)urea (97) from the reaction of S-phenylthioformate with azide ion.

According to the HSAB concept, the failure to prepare sulfonylnitriles by reaction of sulfonyl halides with alkali cyanides (98, 99) is not at all surprising. The alternative type of interaction is highly favored, whereas the desired transformation involves a hard–soft encounter.

$$RSO_2X + CN^{\ominus} \longrightarrow RSO_2^{\ominus} + XCN$$

The reduction of benzenesulfonyl chloride to diphenyl disulfide by triphenylphosphine (100) must be initiated by chlorine abstraction.

Tosylates are susceptible to nucleophilic attack at the hard S and the soft C centers. The site is determined by the hardness of the base employed (101).

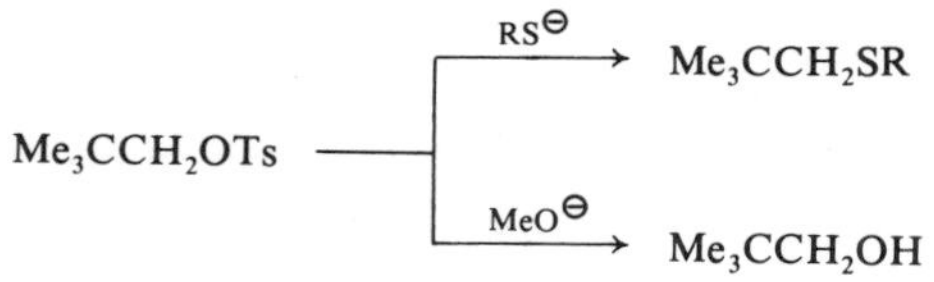

Me_3CCH_2OTs
$\xrightarrow{RS^{\ominus}}$ Me_3CCH_2SR
$\xrightarrow{MeO^{\ominus}}$ Me_3CCH_2OH

4.5.2. Ambident Nucleophiles

Ambident nucleophilic reactivity is more intriguing and complex, as the reaction is influenced by numerous factors—countercation, electrophile, solvent, temperature, and the inherent structural characteristics of the nucleophile. The general behavior of ambident species has been summarized (102). Recently, Gompper and Wagner (102a) developed the concept of allopolarization in an attempt to describe substituent effects on reactions of ambident anions.

Kornblum's work on ambident reactivity is classic. He analyzed a number of systems carefully and laid the foundation of present-day knowledge in this area.

4.5.2.1. N vs. O

The reaction of benzyl bromides with silver nitrite (72) has been studied. The ratio of the products, benzyl nitrite and α-nitrotoluenes, as well as the reaction rates, are intimately related to the structure of the benzyl bromide. The softer electrophiles tend to react preferentially with the softer nitrogen of the base.

Simple nonconjugate alkyl halides participate in the reaction according to the same pattern. The "naked" nitrite ion reacts with alkyl halides giving only nitroalkanes (103). On the contrary, the use of AgNO$_2$ usually leads to nitrite esters as the softer nitrogen center is tied up by Ag$^\oplus$. The cation also assists ionization of the alkylating agents when the latter are halides.

$$Me_3CCl + NO_2^\ominus \longrightarrow Me_3C-ONO + Cl^\ominus$$

$$MeI + NO_2^\ominus \longrightarrow MeNO_2 + Cl^\ominus$$

Reaction of nitramine anions $RNNO_2^\ominus$ with benzyl halides occurs predominantly at nitrogen. The O/N product ratio is enhanced to approximately unity when harder electrophiles (e.g., EtOCH$_2$Cl) are the reaction partners (104).

There has been a controversy concerning the protonation site of amides (105). From the HSAB point of view O-protonation is kinetically favored. Carbamates undergo O-protonation only.

Formamide reacts with alkyl halides to furnish N-alkyl derivatives. Attack by the amidic oxygen on harder acceptors such as α-halo acids and esters has been revealed (106). Secondary amides undergo O-methylation with methyl fluorosulfate (107, 108).

Carbamates afford N-methylated products under equilibrium conditions (109). However, it has been demonstrated that O-methylation by MeOSO$_2$F is rapid and the initial products tend to isomerize to the quaternary salts (110).

Enamides are ambident at C, N, and O, and accordingly they exhibit characteristic alkylation patterns (111). Selective C-methylation of the silver salt with MeI occurs because the vinylidene carbon is soft and the nitrogen site is blocked by its close association with the counterion $Ag^{\oplus}$.

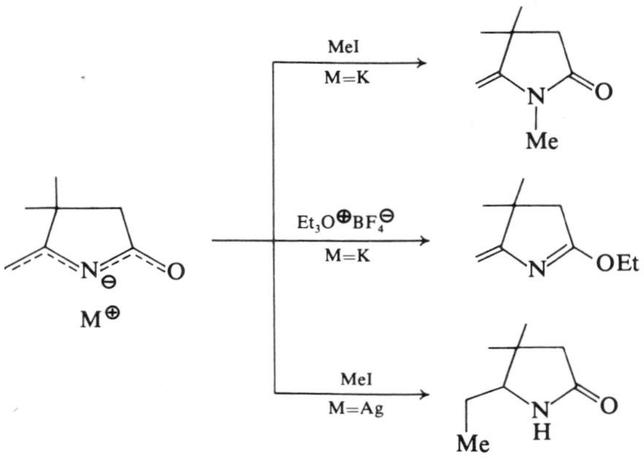

Imides are alkylated with alkyl halides at N only (112). The formation of *N*-alkylimides (113) by treatment of the lithium salts with alkyl chloroformates actually involves O-acylation, which is followed by decarboxylative alkyl transfer from O to N.

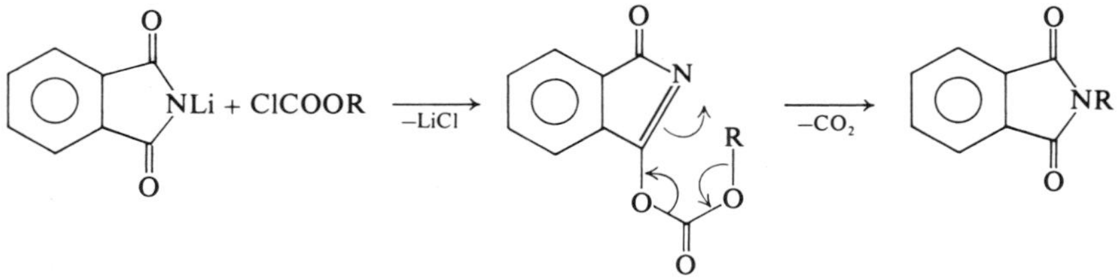

Sodium 9-fluorenone oximate, as dissociated ions (114), is alkylated to the extent of 65% at O and 35% at N. When existing as ion pairs the oximate reacts to give 30% *O*-methyl product. In the presence of a crown ether the oximate ion undergoes almost exclusive (95–99%) O-methylation with MeOTs and O/N-methylation (65:35) with MeI (115). Addition of $NaBPh_4$ to suppress dissociation of the sodium oximate reduces the alkylation with both MeOTs and MeI to nearly the same rate and gives essentially an identical proportion (*ca.* 43:57) of O/N-alkylation. Under such conditions, the two heteroatoms are of approximately the same hardness.

The effect of increased pressure is manifested in a higher O/N ratio. This is due to a shift in the ion pair–free ion equilibrium to the latter (116).

There are three nucleophilic heteroatoms in the amide oximes which can give rise to three types of alkylation products. The course of methylation can be followed by nmr. Thus, it has been shown that only O and N_{imino} products are generated on mixing α-amino-p-tolualdoxime with various methylating agents in CD_3NO_2 (117). Addition of an acid favors the O-methylation.

The trend is unmistakably clear: The harder methylating agent gives larger amounts of the oxime ether by reacting at the oxygen terminus.

Hydroxamic acids exist in two tautomeric modifications of N-hydroxylamide and N-hydroxyimino acid (118). The phenomenon occurs for N,O-diacylhydroxamines also.

The silver or alkali salts of N,O-dibenzoyl derivatives are alkylated at oxygen; O-benzyl benzohydroxamate is derivatized at nitrogen in the presence of NaOMe (119). It is of interest to note that lengthening of the chain in the alkylator leads to a gradual increase in O-alkylation: MeI (100 N), EtI (100 N), n-PrI (75 N, 25 O), n-BuI (70 N, 30 O). Exclusive N-methylation also occurs with MeI, but the O-methyl derivative is evident when Me_2SO_4 is used. Diazomethane effects O-methylation of O-alkyl hydroxamates (120).

In general, O-alkylation can be enhanced at the expense of N-alkylation by using dipolar aprotic solvents, and only O-alkylation is observed in a heterogeneous state (Ag salt, RCH_2X in ether) (121).

Benzohydroxamic acid is less reactive than its N-methyl derivative toward alkyl tosylates (122) because $PhCON(Me)\,O^{\ominus}$ is harder than $PhC=NOH$.
$$\underset{O^{\ominus}}{|}$$

Addition of N,N-dichlorobenzenesulfonamide to 2-butene gives two products (123). The softer nitrogen end of the chlorosulfonamide anion seems to be more apt to approach the incipient carbenium center which is moderately hard.

major   minor

4.5.2.2. N vs. C

Enamines (124, 125) are N,C-ambident. Interestingly, their direct C-protonation with carboxylic acids and N-protonation with HCl have been observed (126). These results can be explained in terms of the HSAB theory (127).

N-Methylation of 1-pyrrolidinocyclohexene is increased (110) fivefold by changing the electrophile from MeI to $MeOSO_2F$. In general, the enamine alkylation is particularly amenable to chain lengthening at the carbon terminus by using soft acceptors such as α,β-unsaturated esters, nitriles, and ketones (124).

Reaction of enamines with acyl halides followed by hydrolysis leads to β-diketones. The apparent violation of the HSAB principle is due to the lack of a forward path for the N-acyl intermediates. C-Acylation is the only feasible reaction course.

Arylation of enamines with activated aromatic halides (128) is possible. It is of interest to note, however, that less reactive (harder) aryl halides could effect substitution at the nitrogen of enamines.

Other soft electrophiles such as benzyne (128) and diazonium ions (128) are avid reaction partners of enamines. α-Arylhydrazonation of ketones results from the latter process.

α-Cyanation of ketones can be achieved in a one-pot reaction via enamines (129). The electrophile is cyanogen chloride. The analogous reaction with cyanogen bromide (130) takes a different course giving α-bromoketones.

The incongruous patterns can be explained with the aid of the HSAB principle. From polarizability measurements (131, 132), it may be concluded that the softness scale is $Br^{\oplus} > CN^{\oplus} > Cl^{\oplus}$. The soft carbon nucleophile of the enamine systems seeks out the softer acid.

Alkyl groups may be introduced into the β position of indole by interacting the halomagnesium salts of the latter with alkyl halides. N-Alkylation occurs by merely changing the countercation to the harder $Na^{\oplus}$ or $K^{\oplus}$. Treatment of alkali salts of 2-phenylindole with dibenzoyl peroxide (133) furnishes only the 3-benzoyloxy derivative. The reaction course is thus completely governed by the soft acceptor characteristic of the peroxide.

The pyrrole anion behaves similarly. Moreover, the fraction of N-allylation product increases when allyl tosylate is used instead of the softer allyl bromide (134).

The ambident 4-pyridylmethide ion can be trapped on the nitrogen with ethyl chloroformate (135). The homolog ion forms spiro derivatives on treatment with hard organohalogen compounds. However, it is not known whether C-alkylation would occur with the soft alkyl halides.

Imine anions attack alkylating agents (136) in accordance with HSAB principle. The selectivity is very similar to that exhibited by the isoelectronic enolate ions.

$$\underset{\text{NPh}}{\overset{\text{Ph}\diagdown\diagup\text{Me}}{\bigg|\bigg|}} \xrightarrow[\text{RX}]{\text{NaH}} \underset{\text{EtNPh}}{\overset{\text{Ph}\diagdown\diagup\text{CH}_2}{\bigg|\bigg|}} + \underset{\text{NPh}}{\overset{\text{Ph}\diagdown\diagup\text{CH}_2\text{Et}}{\bigg|\bigg|}}$$

$RX = EtI$	0.1	1
$RX = Et_2SO_4$	1.2	1
$RX = Et_3O^{\oplus}BF_4^{\ominus}$	22	1

Reaction of phenylhydrazone and oxime dianions with one equivalent of alkyl halide (137) takes place at the carbon.

α-Cyano carbanions undergo C-alkylation with alkyl halides, but they give solely ketenimine derivatives on reaction with the hard trialkylsilyl chlorides (138).

$$\underset{M^{\oplus}}{\overset{\ominus}{RCHCN}} \xrightarrow{R'X} RR'CHCN + MX$$

$$\underset{M^{\oplus}}{\overset{\ominus}{RCHCN}} \xrightarrow{R'_3SiCl} RCH{=}C{=}NSiR'_3 + MCl$$

The cyanide ion is C, N-ambident. Alkali cyanides are extensively used in effecting transformation of alkyl halides into nitriles. Isonitrile formation is promoted by using $Ag^{\oplus}$ which polarizes the C–halogen bond rendering it more ionic (harder).

Both trityl chloride and perchlorate furnish the same product ratio reaction with tetraphenylarsonium cyanide (139), indicating that the electrophile must be the same—trityl cation. The preponderance of nitrile over isonitrile (9 : 1) must mean that the cation is relatively soft.

$$Ph_3C{-}X + Ph_4As^{\oplus}\,CN^{\ominus} \xrightarrow[-Ph_4AsX]{} Ph_3C{-}CN + Ph_3C{-}NC$$

4.5.2.3. O vs. C

α-Diazoketones are protonated at the oxygen (140) when dissolved in the FSO_3H–SbF_5–SO_2 mixture. The previous report claiming C–H bond formation is thereby refuted.

Enolates of carbonyl compounds are ambident anions. Although the negative charge resides predominantly at the oxygen, reactions of enolates with electrophiles can take place at the carbon terminus. C-Monoalkylation of enolates is usually accompanied by di- and polyalkylation owing to a rapid equilibration of the starting enolates with the monoalkylated compounds. The equilibration diminishes with enhancing covalent character of the O–metal bond in the enolates, for example, lithium enolates show a lesser tendency to equilibrate than the corresponding $Na^\oplus$ and $K^\oplus$ salts. In terms of the HSAB principle covalency imparts softness to the oxyanions resulting in a lesser eagerness to abstract the hard proton.

Addition of Bu_3SnCl or Et_3Al to the enolate solutions prior to alkylation suppresses polyalkylation (141). Ketone enolates generated by cleavage of tributyltin enol ethers (142) appear to behave analogously.

Coates *et al.* (143) have developed an elegant procedure for selective quaternization of the carbon alpha to a ketone group. For example, cyclohexanone may be sulfenylated and the substituent then serves as an activator for alkylation and as a blocking group. Regiospecific generation of an enolate with concomitant desulfurization is achieved by lithium in liquid ammonia; the enolate is then alkylated for the second time. Note that every step involves soft–soft interaction.

On the other hand, enolates almost always form trialkyl enol ethers on reaction with the hard chlorosilanes (144). Ethylation of 4-*tert*-butylcyclohexanone lithium enolate (145) with iodoethane leads solely to α-ethylated ketones. However, up to 17% enol ether is obtained when the Meerwein reagent is used. Acetophenone anion is alkylated at oxygen and carbon in the following ratios—0.1, 3.5, and 4.9—depending on whether the reagent is EtI, Me_2SO_4, or $Et_3O^\oplus$ $BF_4^\ominus$, respectively (146). Amylation of butyrophenone enolate in dimethyl sulfoxide (DMSO) with reference to halide variation has also been studied (147). The lesser amount of enol ether formed corresponds to the softer halide.

PhC=CHEt $\xrightarrow[-NaX]{n\text{-}AmX}$	PhC=CHEt	+	PhC—CHEt
$\overset{\mid}{\underset{Na^\oplus}{O^\ominus}}$	$\overset{\mid}{OAm}$		$\overset{\parallel}{O}\ \overset{\mid}{Am}$
X=Cl	1.2		1
X=Br	0.64		1
X=I	0.23		1

Solvent effects play an important role in the alkylation of 9-benzoylfluorene (148). In protic solvents C-alkylation predominates; O-alkylation becomes significant in HMPT, presumably owing to the strong solvation of the cations.

This phenomenon is discernible also during acylation of ketone enolates. The results of acylation of the bromomagnesium salt, generated by conjugate addition of Grignard reagent to benzalacetone (149), indicate a great enhancement of the O-alkylation product fraction. This is accompanied by depletion of the C-alkylated compound when HMPT is added.

$$\underset{\underset{\text{OMgBr}}{|}}{\text{Ph}_2\text{CHCH}}\!\!=\!\!\underset{\text{Me}}{\diagdown} \longrightarrow \underset{\underset{\text{OCOR}}{|}}{\text{Ph}_2\text{CHCH}}\!\!=\!\!\underset{\text{Me}}{\diagdown} + \text{Ph}_2\text{CHCH}\underset{\diagdown\text{COR}}{\overset{\diagup\text{COMe}}{}}$$

Anthrone shows great selectivity in its alkylation (150, 151) producing 9-alkoxyanthracenes with alkyl sulfonates, on the one hand, and 10-alkyl- and 10,10-dialkylanthrones with alkyl halides, on the other. Although benzoins form enediol derivatives readily (e.g., with chloral), exclusive C-alkylation with alkyl halides is observed (152).

Of special interest is the reaction of benzophenone dianion $\text{Ph}_2\text{C}^\ominus\text{–O}^\ominus$ with MeX (153). Despite the unusual disposition of two negative charges on adjacent atoms, no exception to the general rule of leaving group variation with respect to reaction site is noted. The C/O ratios for a few methylating agents have been determined: 7.7 (MeI), 4.2 (MeBr), 2.7 (Me_2SO_4), and 0 (MeOTs).

The α-carbanions of esters are more basic (harder), and exclusive C–C bond formation occurs with MeI, Me_2SO_4, MeOSO_2F, and $\text{Et}_3\text{O}^\oplus\text{BF}_4^\ominus$. Methyl 9-fluorenecarboxylate α-carbanion is aminated by 2,4-dinitrophenoxyamine (154).

As α-alkylated products are thermodynamically more stable, it is significant to note that these ester carbanions undergo both C-silylation and enol silylation (155); the O,α-dianions of carboxylic acids furnish bistrimethylsilyl enol ethers (156).

Disubstituted malonic esters are dealkoxycarbonylated under the modified acyloin condensation condition (Na–Me_3SiCl). Thus, the intermediary ester enolates are also alkylated at the oxygen with the hard acid.

$$\underset{\underset{\text{R}}{\overset{\text{R}}{}}}{\diagup}\!\!\!\!\!\underset{\text{COOMe}}{\overset{\text{COOMe}}{\diagdown}} \longrightarrow \underset{\underset{\text{R}}{\overset{\text{R}}{}}}{}\diagdown\!\!=\!\!\diagup\underset{\text{OSiMe}_3}{\overset{\text{OMe}}{}}$$

Interestingly, trimethylsilylation of *tert*-butyl lithio(α-trimethylsilyl)acetate occurs predominantly at carbon (157). Symbiosis favors the observed reaction pattern.

Considerable attention has been paid to the elucidation of the ambient behavior of β-dicarbonyl compounds. Pearson's principle proves valuable for rationalizing various observations. Data are available concerning the alkylation of β-keto esters and β-diketones as a function of alkyl halides. Ethyl sodioacetoacetate gives yields of O-methylated product (158) in DMSO of 6, 26, and 43% with the iodide, bromide, and chloride, respectively. The increase of O-alkylation with the branching of the halogen-bearing carbon atom (EtBr, 26%; *n*-PrBr, 47%; *i*-PrBr, 62%) agrees with a progressively harder electrophilic center.

Reutov and co-workers (159–161) have found that combining substituted acetoacetate enolates ($RCOC^{\ominus}HCOOEt$ $M^{\oplus}$) with ethyl tosylate in HMPT furnishes C and O ethylated products whose ratio is independent of the counterion (Li, Na, K, or Cs). Although both products increase with inductive electron release by the R group, only O-alkylation is observed when R = CF_3. Variation of k_c/k_0 according to leaving group is in the order I > Br > Cl > OTs, and it is very large for MeX, but negligible for sec-BuX. As X = OTs → X = I, k_c increases by 10^3, whereas k_0 changes by a factor of 10 only.

Such are excellent demonstrations of the HSAB principle since soft–soft interactions are particularly sensitive to subtle changes in softness of the reaction partners as contrasted to hard–hard interactions in which the overall reaction is dominated by electrostatic forces.

Exclusive O-alkylation (162, 163) of β-dicarbonyl compounds takes place with $ClCH_2OR$ which resembles acyl halides (164–166). Transition state symbiosis contributes to the hard-type reaction. On the other hand, enolates attack the much softer $ClCH_2SMe$ with their carbon termini (167).

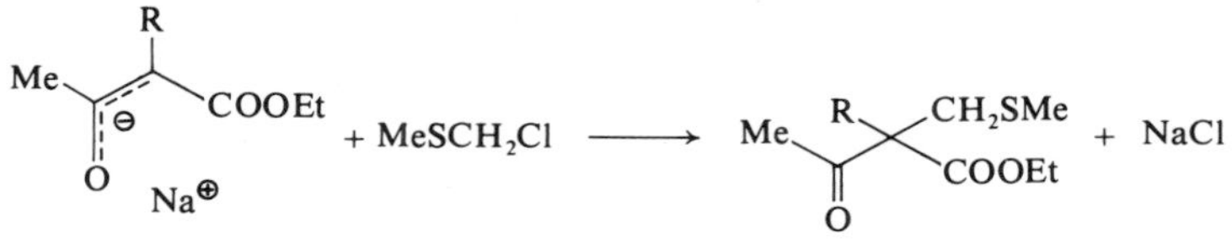

There is no apparent relationship between reaction rates and the C/O product ratio in the ethylation of acetoacetate (168). The k_c/k_0 trend is I > Br $\gg$ TsO > $EtSO_4$ > CF_3SO_3 $\simeq$ FSO_3, while the overall rate constants vary irregularly with respect to the softness of the leaving groups.

Sodioacetoacetate reacts with $Et_3O^{\oplus}$ $BF_4^{\ominus}$ to give cis-O-ethyl enol ether (169).

Results from alkylation of acetylacetone enolates in dipolar aprotic solvents (170) parallel closely those of ethyl acetoacetate. Zaugg et al. (171) reported the interaction of a hindered β-diketone enolate, sodiodipivaloylmethane, with trityl halides. O-Alkylation is more important with the chloride than with the bromide as usual. The striking feature of the reaction is that the carbanion attacks the aryl carbon of the electrophile as a consequence of steric repulsion.

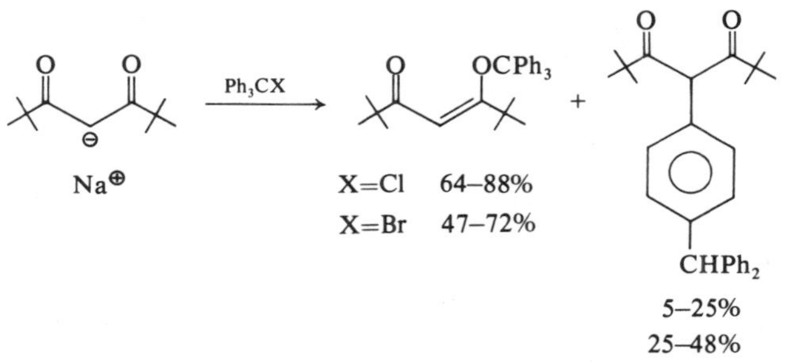

Acylation is charge-controlled. The variation in hardness of the acylating agents is often reflected in a corresponding shift in product ratio, as demonstrated by the increasing tendency of 2-carbethoxycyclanones toward O-acylation (172) with $ClCOOEt < ClCOCH_2Cl < AcCl < AcClO_4$.

The phenyliodonium dimedone ylide attacks diphenylketene with the oxygen atom (173). Interestingly, the reaction of the ylide with thiourea (174) occurs at C-2, presumably via the following mechanism.

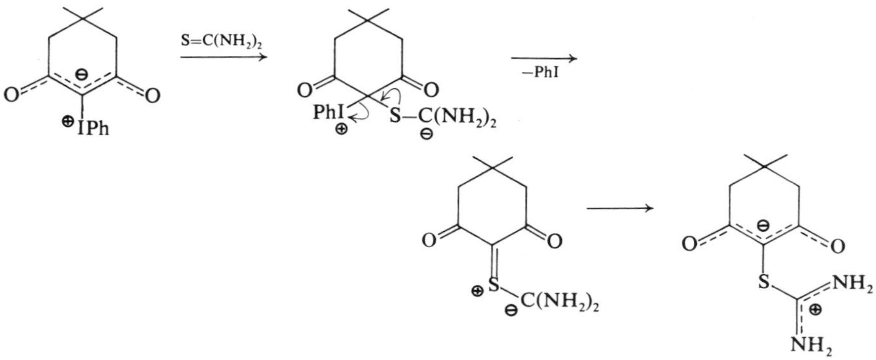

Sulfonylation of aroylmalonates gives enol derivatives (175) because sulfonyl chlorides are even harder than acyl chlorides. The α-alkoxy-β-oxosulfone anions form sulfonyl reductone derivatives on reaction with $ClCH_2OMe$ or aroyl chlorides (163). This behavior is in direct contrast to the C-alkylation with alkyl halides.

Stabilized ylides are acylated at oxygen (176, 177) in accordance with HSAB theory.

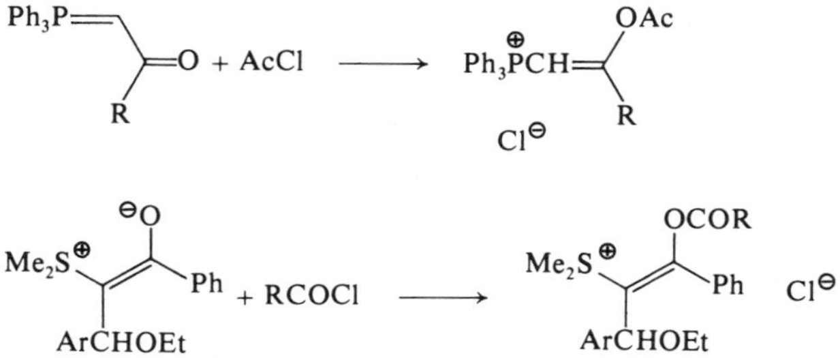

The study of phenol alkylation has a long history. Claisen and co-workers (178) showed that metal phenoxides react with active alkyl halides (e.g., allyl and benzyl bromides) in nonpolar solvents to give o-alkylphenols. Cyclohexadienones may be prepared in this way starting from 2,6-disubstituted phenols

(179). Kornblum and Lurie (180) suggested that only C-alkylation occurs under completely heterogeneous conditions and only O-alkylation in homogeneous media. This statement stands as a guide in synthetic planning.

Correlation of phenolate reaction sites with electrophiles is as valid as in the other ambident systems. Thus, only the methyl ethers are produced from the reaction of sodium 2,6-dimethylphenoxide with methyl brosylate (181) or sodium 2-naphthoxide with dimethyl sulfate (182). Analysis of aroylation of 2-naphthoxides in hexane (183) indicates that (*i*) softer Li enolate gives higher C/O ratios than the Na enolate, and (*ii*) more C-acylation occurs with the aroyl chloride of softer nature.

The thallous salt of 2-hydroxythiophene is methylated (184) mainly at C-3 with MeI.

2,6-Di-*tert*-butyl-4-nitrophenol is methylated by CH_2N_2 and silylated by $ClSiMe_3$ to give nitronate esters. Apparently steric crowding around the phenolic group forces the electrophiles to react with the alternative hard basic center. The less hindered 2,6-diisopropyl-4-nitrophenol affords a mixture of methyl ether and nitronate ester (185, 186).

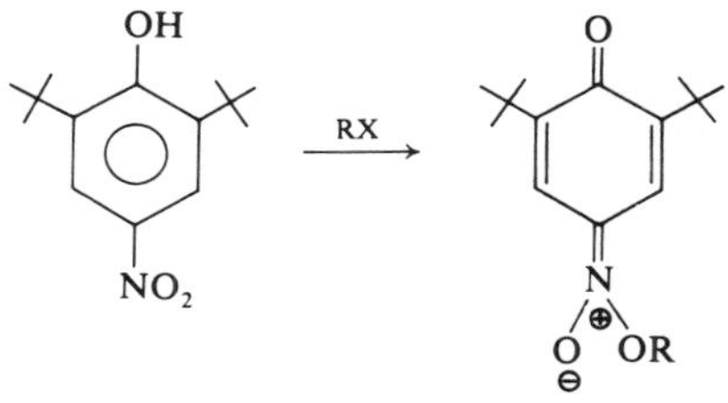

4.5.2.4. Others

Unsymmetrically substituted allyl carbanions are ambident. The reaction of phenylated allyl carbanions with various methylating agents has been investigated (187). The reactivity pattern corresponds with leaving group hardness.

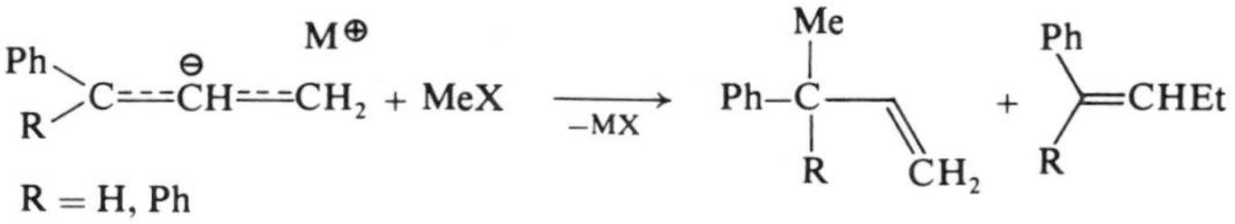

1,1-Dichloroallyllithium adds to ordinary ketones with its softer CCl_2 terminus, but to aryl ketones and hexafluoroacetone with its harder CH_2 end (188).

The hard oxygen of hydroxyphosphines is readily acylated, while alkylation takes place preferentially at the phosphorus (189).

REFERENCES

1. C. K. Ingold, "Structure and Mechanism in Organic Chemistry," 2nd ed. Cornell Univ. Press, Ithaca, New York, 1969.
2. S. Winstein, E. Grunwald, and H. W. Jones, *J. Am. Chem. Soc.* **73**, 2700 (1951).
3. A. Streitwieser, *Chem. Rev.* **56**, 571 (1956).
4. J. C. Harris and J. L. Kurz, *J. Am. Chem. Soc.* **92**, 349 (1970).
5. J. Cantacuzène and J. Leroy, *Tetrahedron Lett.* p. 3277 (1970).
6. A. Rosowsky, *in* "Heterocyclic Compounds with Three- and Four-Membered Rings" (A. Weissberger, ed.), Part 1, pp. 316–327. Wiley (Interscience), New York, 1964.
7. R. Luhowy and F. Meneghini, *J. Org. Chem.* **38**, 2405 (1973).
8. M. D. Bentley, I. B. Douglass, J. A. Lacadie, D. C. Weaver, F. A. Davis, and S. J. Eitelman, *Chem. Commun.* p. 1625 (1971).
9. J. F. Bunnett and R. E. Zahler, *Chem. Rev.* **49**, 273 (1951).
10. C. G. Swain and C. B. Scott, *J. Am. Chem. Soc.* **75**, 141 (1953).
11. A. J. Parker, *Proc. Chem. Soc. London*, p. 371 (1961).
12. J. O. Edwards, *J. Am. Chem. Soc.* **76**, 1540 (1954).
13. J. O. Edwards, *J. Am. Chem. Soc.* **78**, 1819 (1956).
14. J. O. Edwards and R. G. Pearson, *J. Am. Chem. Soc.* **84**, 16 (1961).
15. J. F. Bunnett, *Annu. Rev. Phys. Chem.* **14**, 271 (1963).
16. J. F. Bunnett and J. D. Reinheimer, *J. Am. Chem. Soc.* **84**, 3284 (1962).
17. R. F. Hudson and G. Klopman, *Helv. Chim. Acta* **44**, 1914 (1961).
18. S. Winstein, L. G. Savedoff, G. Smith, I. D. Stevens, and J. S. Gall, *Tetrahedron Lett.* No. 9, p. 24 (1960).
19. J. E. Gordon and P. Varughese, *Chem. Commun.* p. 1160 (1971). cf. C. Minot and Nguyen Trong Anh, *Tetrahedron Lett.* p. 3905 (1975).
20. J. Dockx, *Synthesis* p. 411 (1973).
21. A. Merz and G. Märkl, *Angew. Chem., Int. Ed. Engl.* **12**, 845 (1973); A. Merz, *ibid.* p. 846.
22. G. N. Schrauzer, *Acc. Chem. Res.* **1**, 97 (1968).
23. G. N. Schrauzer, E. Deutsch, and R. J. Windgassen, *J. Am. Chem. Soc.* **90**, 2441 (1968).
24. R. Grigg, A. W. Johnson, and G. Shelton, *Chem. Commun.* p. 1151 (1968).
25. T. C. Bruice and R. Lapinski, *J. Am. Chem. Soc.* **80**, 2265 (1958).
26. W. P. Jencks and J. Carriuolo, *J. Am. Chem. Soc.* **82**, 1778 (1960).
27. R. F. Hudson and M. Green, *J. Chem. Soc.* p. 1055 (1962).
28. L. R. Fedor, T. C. Bruice, K. L. Kirk, and J. Meinwald, *J. Am. Chem. Soc.* **88**, 108 (1966).
29. J. Grundy, B. G. James, and G. Pattenden, *Tetrahedron Lett.* p. 157 (1972).
30. T. Cohen and A. H. Lewin, *J. Am. Chem. Soc.* **88**, 4521 (1966).
31. A. H. Lewin and N. L. Goldberg, *Tetrahedron Lett.* p. 491 (1972).
32. T. Saegusa and I. Murase, *Synth. Commun.* **2**, 1 (1972).
33. T. A. Bryson, *Synth. Commun.* **2**, 361 (1972).
34. R. C. Larock, *J. Org. Chem.* **39**, 3721 (1974).
35. H. Normant and T. Cuvigny, *Bull. Soc. Chim. Fr.* p. 1867 (1965).
36. J. E. Shaw, D. C. Kunerth, and J. J. Sherry, *Tetrahedron Lett.* p. 689 (1972).
37. P. E. Pfeffer, T. A. Foglia, P. A. Barr, I. Schmeltz, and L. S. Silbert, *Tetrahedron Lett.* p. 4063 (1972).
38. R. C. Larock, *J. Org. Chem.* **39**, 3721 (1974).
39. J. C. Sheehan and G. D. Daves, Jr., *J. Org. Chem.* **29**, 2006 (1964).
40. N. Kornblum and A. Scott, *J. Am. Chem. Soc.* **96**, 590 (1974).

41. P. A. Bartlett and W. S. Johnson, *Tetrahedron Lett.* p. 4459 (1970).
42. T.-L. Ho and C. M. Wong, *Synth. Commun.* **4**, 307 (1974).
43. T.-L. Ho, *Synthesis* p. 715 (1974).
44. T.-L. Ho, *Synthesis* p. 510 (1975).
45. A. C. Pierce and M. M. Joullie, *J. Org. Chem.* **27**, 3968 (1962).
46. T.-L. Ho and C. M. Wong, *Synth. Commun.* **5**, 305 (1975).
47. E. W. Thomas and T. I. Crowell, *J. Org. Chem.* **37**, 744 (1972).
48. E. Taschner and B. Liberek, *Rocz. Chem.* **30**, 323 (1956).
49. F. Elsinger, J. Schreiber, and A. Eschenmoser, *Helv. Chim. Acta* **43**, 113 (1960).
50. P. D. G. Dean, *J. Chem. Soc.* p. 6655 (1965).
51. J. E. McMurry and G. B. Wong, *Synth. Commun.* **2**, 389 (1972).
52. B. Saville, *Angew. Chem., Int. Ed. Engl.* **6**, 928 (1967).
53. J. Halpern, *Q. Rev., Chem. Soc.* **10**, 463 (1956); *Annu. Rev. Phys. Chem.* **16**, 103 (1965).
54. M. T. Beck, I. Gimesi, and J. Farkas, *Nature (London)* **197**, 73 (1963).
55. R. E. Dessy and F. E. Paulik, *J. Am. Chem. Soc.* **85**, 1812 (1963).
56. H. A. Hageman, *Org. React.* **7**, 198 (1953).
57. P. Rumpf, *C. R. Hebd. Seances Acad. Sci., Ser. C* **198**, 256 (1934).
58. H. A. Liebhafsky and A. Mohammad, *J. Phys. Chem.* **38**, 857 (1934).
59. S. Zeisel, *Monatsh. Chem.* **6**, 989 (1885).
60. H. Pommer, *Angew. Chem.* **72**, 911 (1960).
61. R. L. Frank and P. V. Smith, *J. Am. Chem. Soc.* **68**, 2103 (1946).
62. W. P. Jencks and J. Carriuolo, *J. Biol. Chem.* **234**, 1272 and 1280 (1959).
63. F. Covitz and F. H. Westheimer, *J. Am. Chem. Soc.* **85**, 1773 (1963).
64. C. N. Lieske, J. W. Hovanec, and P. Blumbergs, *Chem. Commun.* p. 976 (1969).
65. D. Martin, K. Nadolski, R. Bacaloglu, and I. Bacaloglu, *J. Prakt. Chem.* **313**, 58 (1971).
66. H. T. Openshaw and N. Whittaker, *J. Chem. Soc. C* p. 89 (1969).
67. C. G. Swain and J. F. Brown, Jr., *J. Am. Chem. Soc.* **74**, 2538 (1962).
68. P. R. Rony, *J. Am. Chem. Soc.* **91**, 6090 (1969).
69. D. Bryce-Smith, *Proc. Chem. Soc., London* p. 20 (1957).
70. U. Schöllkopf and N. Rieber, *Chem. Ber.* **102**, 488 (1969).
71. H. Rheinboldt and E. Motzkus, *Chem. Ber.* **72**, 657 (1939).
72. N. Kornblum, R. A. Smiley, R. K. Blackwood, and D. C. Iffland, *J. Am. Chem. Soc.* **77**, 6269 (1955).
73. S. Hünig, *Angew. Chem., Int. Ed. Engl.* **3**, 548 (1964).
74. A. Holy, *Tetrahedron Lett.* p. 585 (1972).
75. W. P. Tucker and G. L. Roof, *Tetrahedron Lett.* p. 2747 (1967).
76. T. Nakai and M. Okawara, *Bull. Chem. Soc. Jpn.* **43**, 1864 (1970).
77. A. Schmidpeter, *Tetrahedron Lett.* p. 1421 (1963).
78. S. Hünig and H. Hansen, *Chem. Ber.* **102**, 2109 (1969).
79. S. Hünig and W. Kniese, *Justus Liebigs Ann. Chem.* **708**, 170 (1967).
80. W. H. Richardson and V. F. Hodge, *Tetrahedron Lett.* p. 749 (1971).
81. W. H. Richardson and V. F. Hodge, *J. Am. Chem. Soc.* **93**, 3996 (1971).
82. T. L. Gresham, J. E. Jansen, F. W. Shaver, J. T. Gregory, and W. L. Beears, *J. Am. Chem. Soc.* **70**, 1004 (1948).
83. T. L. Gresham, J. E. Jansen, F. W. Shaver, M. R. Frederick, F. T. Fiedorek, R. A. Bankert, J. T. Gregory, and W. L. Beears, *J. Am. Chem. Soc.* **74**, 1323 (1952).
84. T. L. Gresham, J. E. Jansen, F. W. Shaver, and J. T. Gregory, *J. Am. Chem. Soc.* **70**, 999 (1948).
85. J. Harley-Mason, *J. Chem. Soc.* p. 2433 (1952).
86. K. Itoh, Y. Kato, and Y. Ishii, *J. Org. Chem.* **34**, 459 (1969).
87. C. Couret, J. Escudié, and J. Satgé, *Recl. Trav. Chim. Pays-Bas* **91**, 429 (1972).

88. G. Guanti, C. Dell'Erba, F. Pero, and G. Leandri, *Chem. Commun.* p. 823 (1975).
89. A. Hassner and A. Kascheres, *Tetrahedron Lett.* p. 4623 (1970).
90. D. Seebach and D. Enders, *Angew. Chem., Int. Ed. Engl.* **11**, 301 (1972).
91. P. Buck and G. Köbrich, *Tetrahedron Lett.* p. 1563 (1967).
92. F. L. Scott, R. E. Oesterling, E. A. Tyczkowski, and C. E. Inman, *Experientia* **14**, 355 (1958).
93. C. E. Inman, R. E. Oesterling, and E. A. Tyczkowski, *J. Am. Chem. Soc.* **80**, 6533 (1958).
94. E. Buncel and A. Raoult, *Chem. Commun.* p. 210 (1973).
95. R. Graf, *Justus Liebigs Ann. Chem.* **661**, 111 (1963).
96. H. Vorbrüggen and K. Krolikiewicz, *Chem. Ber.* **108**, 2137 (1975).
97. M. Prince and C. M. Orlando, Jr., *Chem. Commun.* p. 818 (1967).
98. B. Saville, *Chem. Ind. (London)* p. 660 (1956).
99. F. P. Corson and R. G. Pews, *J. Org. Chem.* **36**, 1654 (1971).
100. L. Horner and H. Nickel, *Justus Liebigs Ann. Chem.* **597**, 20 (1955).
101. J. F. Bunnett, and J. Y. Bassett, *J. Am. Chem. Soc.* **81**, 2104 (1959).
102. R. Gompper, *Angew. Chem., Int. Ed. Engl.* **3**, 560 (1964).
102a. R. Gompper and H.-U. Wagner, *Angew. Chem., Int. Ed. Engl.* **15**, 321 (1976).
103. J. W. Zubrick, B. I. Dunbar, and H. D. Durst, *Tetrahedron Lett.* p. 71 (1975).
104. B. Unterhalt and D. Thamer, *Tetrahedron Lett.* p. 4905 (1971).
105. M. Liler, in *Adv. Phys. Org. Chem.* **11**, 328–351 (1975).
106. H. Bredereck, R. Gompper, and G. Teilig, *Chem. Ber.* **87**, 537 (1954).
107. W. Rundel, *Chem. Ber.* **108**, 1779 (1975).
108. S. Julia and R. J. Ryan, *C. R. Hebd. Seances Acad. Sci., Ser. C* **274**, 1207 (1972).
109. M. G. Ahmed and R. W. Alder, *Chem. Commun.* p. 1389 (1969).
110. R. W. Alder, *Chem. Ind. (London)* p. 983 (1973).
111. A. Eschenmoser, Sandin Lecture, University of Alberta (1967).
112. A. N. Nesmeyanov and M. I. Kabachnik, *Zh. Obshch. Khim.* **25**, 41 (1955).
113. J. A. Vida, *Tetrahedron Lett.* p. 3921 (1972).
114. S. G. Smith and D. V. Milligan, *J. Am. Chem. Soc.* **90**, 2393 (1968).
115. S. G. Smith and M. P. Hanson, *J. Org. Chem.* **36**, 1931 (1971).
116. W. J. LeNoble and S. K. Palit, *Tetrahedron Lett.* p. 493 (1972).
117. H. Gozlan, R. Michelot, and R. Rips, *C. R. Hebd. Seances Acad. Sci., Ser. C* **278**, 363 (1974).
118. W. Lossen, *Justus Liebigs Ann. Chem.* **252**, 170 (1889); **281**, 169 (1894).
119. M. Chehata, F. Bocabeille, G. Thuillier, and P. Rumpf, *C. R. Hebd. Seances Acad. Sci., Ser. C* **268**, 445 (1969).
120. R. Blaser, P. Imfeld, and O. Schindler, *Helv. Chim. Acta* **52**, 569 (1969).
121. J. E. Johnson, J. R. Springfield, J. S. Hwang, L. J. Hayes, W. C. Cunningham, and D. L. McClaugherty, *J. Org. Chem.* **36**, 284 (1971).
122. M. Dessolin, M. Laloi-Diard, and M. Vilkas, *Tetrahedron Lett.* p. 2405 (1974).
123. F. A. Daniher, M. T. Melchior, and P. E. Butler, *Chem. Commun.* p. 931 (1968).
124. G. Stork, A. Brizzolara, H. Landesman, J. Szmuszkovicz, and R. Terrell, *J. Am. Chem. Soc.* **85**, 207 (1963).
125. A. G. Cook, "Enamines: Synthesis, Structure, and Reactions." Dekker, New York, 1969.
126. G. Opitz and A. Griesinger, *Justus Liebigs Ann. Chem.* **665**, 101 (1963).
127. L. Alais, R. Michelot, and B. Tchoubar, *C. R. Hebd. Seances Acad. Sci., Ser. C* **273**, 261 (1971).
128. M. E. Kuehne, *J. Am. Chem. Soc.* **84**, 837 (1962).
129. M. E. Kuehne, *J. Am. Chem. Soc.* **81**, 5400 (1959).
130. F. Fusco, S. Rossi, and G. Bianchetti, *Gazz. Chim. Ital.* **91**, 841 (1961).
131. F. Fairbrother, *J. Chem. Soc.* p. 180 (1950).

132. G. Lord and A. A. Wolf, *J. Chem. Soc.* p. 2546 (1954).
133. T. Nishio, M. Yuyama, and Y. Omote, *Chem. Ind. (London)* p. 480 (1975).
134. E. Papadopoulos and K. Tabello, *J. Org. Chem.* **33**, 1299 (1968).
135. G. Fraenkel and J. W. Cooper, *J. Am. Chem. Soc.* **93**, 7228 (1971).
136. G. J. Heiszwolf and H. Kloosterziel, *Recl. Trav. Chim. Pays-Bas* **89**, 1217 (1970).
137. F. E. Henoch, K. G. Hampton, and C. R. Hauser, *J. Am. Chem. Soc.* **91**, 676 (1969).
138. D. S. Watt, *Synth. Commun.* **4**, 127 (1974).
139. T. Austad, J. Songstad, and L. J. Strangeland, *Acta Chem. Scand.* **25**, 2327 (1971).
140. M. Allard, J. Levisalles, and J. M. Sommer, *Chem. Commun.* p. 1515 (1969).
141. P. A. Tardella, *Tetrahedron Lett.* p. 1117 (1969).
142. M. Pereyre and Y. Odic, *Tetrahedron Lett.* p. 505 (1969).
143. R. M. Coates, H. D. Pigott, and J. Ollinger, *Tetrahedron Lett.* p. 3955 (1974).
144. H. O. House, L. J. Czuba, M. Gall, and H. D. Olmstead, *J. Org. Chem.* **34**, 2324 (1969).
145. H. O. House, B. A. Tenfertiller, and H. D. Olmstead, *J. Org. Chem.* **33**, 935 (1968).
146. G. J. Heiszwolf and H. Kloosterziel, *Recl. Trav. Chim. Pays-Bas* **89**, 1153 (1970).
147. H. D. Zook and J. A. Miller, *J. Org. Chem.* **36**, 1112 (1971).
148. J. P. Boisset, J. Boyer, and J. Rouzaud, *C. R. Hebd. Seances Acad. Sci., Ser. C* **263**, 1253 (1966).
149. P. Angibeaud and M.-J. Lagrange, *C. R. Hebd. Seances Acad. Sci., Ser. C* **272**, 1506 (1971).
150. K. H. Meyer and H. Schlosser, *Justus Liebigs Ann. Chem.* **420**, 126 (1920).
151. E. deB. Barnett, J. W. Cook, and M. A. Matthew, *J. Chem. Soc.* **123**, 1994 (1923).
152. J. H. van de Sande and K. R. Kopecky, *Can. J. Chem.* **47**, 163 (1969).
153. W. S. Murphy and D. J. Buckley, *Tetrahedron Lett.* p. 2975 (1969).
154. T. Sheradsky and Z. Nir, *Tetrahedron Lett.* p. 77 (1969).
155. M. W. Rathke and D. F. Sullivan, *Synth. Commun.* **3**, 67 (1973).
156. Y.-N. Kuo, F. Chen, C. Ainsworth, and J. J. Bloomfield, *Chem. Commun.* p. 136 (1971).
157. S. L. Hartzell and M. W. Rathke, *Tetrahedron Lett.* p. 2737 (1976).
158. W. J. LeNoble and J. E. Puerta, *Tetrahedron Lett.* p. 1087 (1966).
159. A. L. Kurts, A. Macias, I. P. Beletskaya, and O. A. Reutov, *Tetrahedron* **27**, 4759 (1971); *Tetrahedron Lett.* p. 3037 (1971).
160. A. L. Kurts, P. I. Dem'yanov, A. Macias, I. P. Beletskaya, and O. A. Reutov, *Tetrahedron* **27**, 4769 (1971).
161. A. L. Kurts, N. K. Genkina, A. Macias, I. P. Beletskaya, and O. A. Reutov, *Tetrahedron* **27**, 4777 (1971).
162. R. M. Coates and J. E. Shaw, *J. Org. Chem.* **35**, 2601 (1970).
163. K. Schank, H. Hasenfratz, and A. Weber, *Chem. Ber.* **106**, 1107 (1973).
164. K. Yoshida and Y. Yamashita, *Tetrahedron Lett.* p. 693 (1966).
165. J. L. Wong and C. H. Savells, Jr., *Org. Prep. Proced. Int.* **3**, 269 (1971).
166. R. E. Davis, *Tetrahedron Lett.* p. 5201 (1966).
167. H. Böhme and E. Mundlos, *Chem. Ber.* **86**, 1414 (1953).
168. P. Sarthou, F. Guibé, and G. Bram, *Chem. Commun.* p. 377 (1974).
169. T. A. Mastryukova, A. E. Shipov, V. V. Abalyaeva, E. E. Kugutcheva, and M. I. Kabachnik, *Dokl. Akad. Nauk SSSR* **164**, 340 (1965).
170. N. K. Genkina, A. L. Kurts, I. P. Beletskaya, and O. A. Reutov, *Dokl. Akad. Nauk SSSR* **189**, 1252 (1969).
171. H. E. Zaugg, R. J. Michaels, and E. J. Baker, *J. Am. Chem. Soc.* **90**, 3800 (1968).
172. J. Ferris, B. Wright, and C. Crawford, *J. Org. Chem.* **30**, 2367 (1965).
173. G. F. Koser and S.-M. Yu, *J. Org. Chem.* **40**, 1166 (1975).
174. G. F. Koser and S.-M. Yu, *J. Org. Chem.* **41**, 125 (1976).
175. I. Fleming and J. Harley-Mason, *J. Chem. Soc.* p. 4771 (1963).

176. P. A. Chopard, R. J. G. Searle, and F. H. Devitt, *J. Org. Chem.* **30**, 1015 (1965).
177. A. Winkler and J. Gosselck, *Tetrahedron Lett.* p. 4229 (1969).
178. L. Claisen, F. Kremers, F. Roth, and E. Tietze, *Justus Liebigs Ann. Chem.* **442**, 210 (1925).
179. D. Y. Curtin, R. J. Crawford, and M. Wilhelm, *J. Am. Chem. Soc.* **80**, 1391 (1958).
180. N. Kornblum and A. P. Lurie, *J. Am. Chem. Soc.* **81**, 2705 (1959).
181. D. Y. Curtin and R. R. Fraser, *J. Am. Chem. Soc.* **80**, 6016 (1958)
182. E. Wenkert, R. D. Youssefyeh, and R. G. Lewis, *J. Am. Chem. Soc.* **82**, 4675 (1960).
183. P. Haberfield and R. B. Trattner, *Chem. Commun.* p. 1481 (1971).
184. E. B. Pedersen and S.-O. Lawesson, *Tetrahedron* **27**, 3861 (1971).
185. J. S. Meek, J. S. Fowler, P. A. Monroe, and T. J. Clark, *J. Org. Chem.* **33**, 223 (1968).
186. J. S. Meek and J. S. Fowler, *J. Org. Chem.* **33**, 226 (1968).
187. W. S. Murphy, R. Boyce, and E. A. O'Riordan, *Tetrahedron Lett.* p. 4157 (1971).
188. D. Seyferth, G. J. Murphy, and R. A. Woodruff, *J. Am. Chem. Soc.* **96**, 5011 (1974); cf. F. Barbot, C. H. Chan, and P. Miginiac, *Tetrahedron Lett.* p. 2309 (1976).
189. E. I. Grinstein, A. B. Bruker, and L. Z. Soborowskii, *Zh. Obshch. Khim.* **36**, 302 (1966).

5

Alkene Chemistry

5.1. FORMATION OF OLEFINIC LINKAGES BY ELIMINATION REACTIONS. ELIMINATION vs. SUBSTITUTION

The most well-known method for double-bond formation is β elimination (1). In a pyrolytic elimination (2) neither a base nor a solvent is needed. The nucleofugal group acts as an internal base in a cyclic E mechanism leading to cis elimination. Most of these reactions involve proton abstraction by a hard terminal atom except those involving the energetic ylide intermediates.

The Cope elimination (3) demands lower temperatures than the Hofmann degradation. This is presumably due to the hardness of the oxygen base which is further augmented by the neighboring nitrogen. The oxidation of secondary alcohols by nitrosonium ion (4) which is shown below could involve a Cope-type elimination.

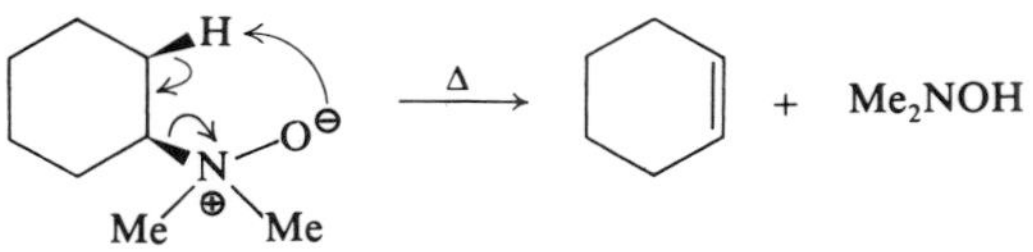

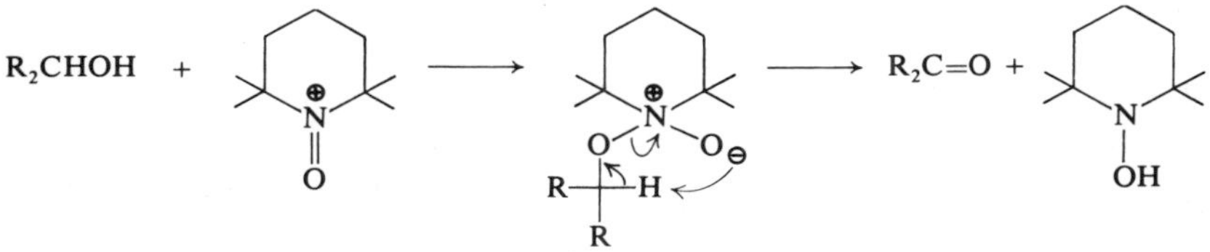

The thermolysis of sulfoxides (5, 6) has gained prominence as a technique for olefin synthesis because of the mild conditions required.

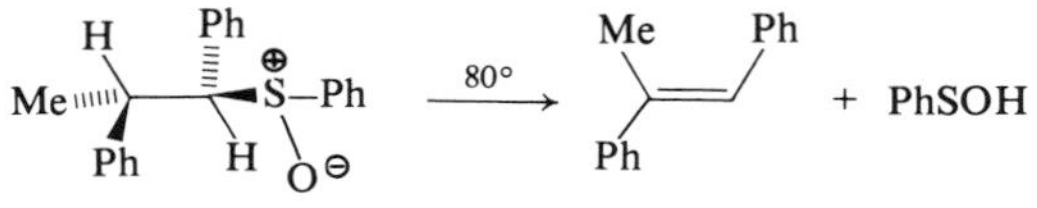

Treatment of alkyl halides with bases can lead to alkenes or substitution products, or both. Hard bases always favor elimination (by attacking the hard β proton), whereas soft donors prefer $S_N 2$ displacement.

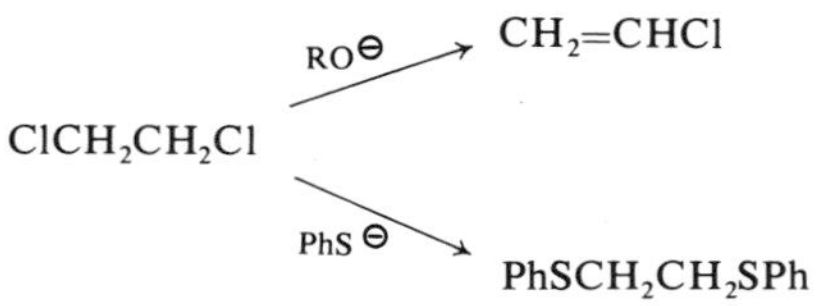

Halide ions, especially $F^{\ominus}$, are weak bases in aqueous solution. However, in dry aprotic solvents $Et_4N^{\oplus}F^{\ominus}$ effectively induces elimination of HBr from 2-bromoethylbenzene (7). The ammonium chloride and bromide only promote halogen exchange. At 80°, the fluoride decomposes into Et_3N, HF, and ethylene, thus demonstrating the high protophilicity of the hard fluoride ion. This same property is also shown by its usefulness for generating dihalocarbenes. "Naked" fluoride ion (cation sequestered by a crown ether) promotes facile elimination of haloethylenes to form acetylenes and allenes (8).

Ethoxide and malonate anions have virtually identical proton basicity (9), but significantly different hardness. As a consequence, ethoxide promotes elimination of 2-bromopropane, whereas malonate effects substitution.

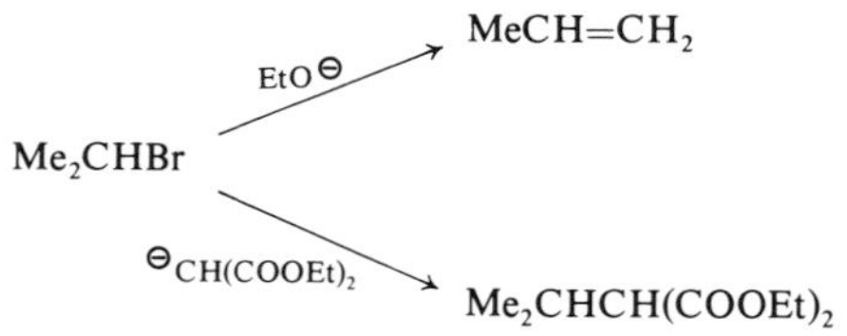

A change in the countercation sometimes has profound influence on the reaction course. N-Sodio-2-methylamino-3,4-dihydronaphthalene induces dehydrohalogenation of 1,2-bromochloroethane, whereas the softer chloromagnesium salt of the same enamine leads to a displacement process (10).

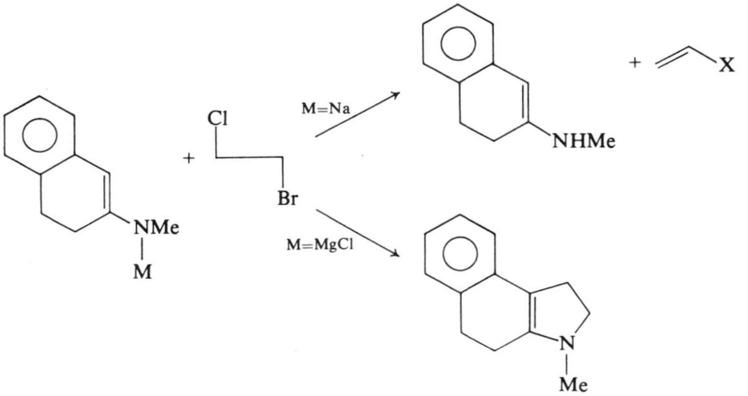

Primary alkyl bromides are dehydrobrominated by *tert*-butoxide ion, but this same base displaces the tosyloxy group from the corresponding alkyl tosylates. The difference must be related to the choice of hard and soft acceptors for the transition symbiosis clearly operates in the latter instances.

The dramatic increase in the $E2/S_N2$ ratio in the reactions of tosylates with oxalate vs. formate anions has been ascribed to a possible bidentate attack on hydrogen by the oxalate ion (11, 12). It should be noted that oxalate is a harder base compared to the formate.

The concerted elimination of vicinal dihalides, in particular, dibromides, may be accomplished by a variety of reagents (13). Both iodide ion (14) and zinc dust which are soft bases are efficient.

The relative halophilicity of trivalent phosphorus compounds, n-$Bu_3P >$ $Ph_3P > (EtO)_3P$, has been demonstrated from their reactions with *vic*-dibromides (15). Their reactivities parallel the softness of the central phosphorus atoms.

The debromination and dehydrobromination of *meso*- and *dl*-stilbene dibromides (16) in dimethylformamide (DMF) have been scrutinized. The protophilicity and bromophilicity of the reagents, $F^\ominus > Cl^\ominus > DMF$ and $I^\ominus > Br^\ominus > Cl^\ominus > Sn^{2\oplus} > DMF$, respectively, are in accordance with the hard and soft scales.

In the 1,3-elimination of bis-α-bromobenzyl sulfones (17), soft bases ($PhS^\ominus$, $I^\ominus$, $H^\ominus$, Ph_3P, Mg, Zn) attack the bromine atom to give stilbenes as the final products, whereas hard bases ($MeO^\ominus$, R_3N, DMF, dimethylacetamide) abstract an α-hydrogen which leads to bromostilbenes (Ramberg–Bäcklund reaction).

Although 9-heterabicyclo[3.3.1]nona-2,6-dienes are the sole products from the double displacement of (Z,Z)-*cis*-3,7-dibromocycloocta-1,5-diene with oxygen, nitrogen, and sulfur bases (18), the very soft sodium telluride also effects a concurrent cycloelimination (19).

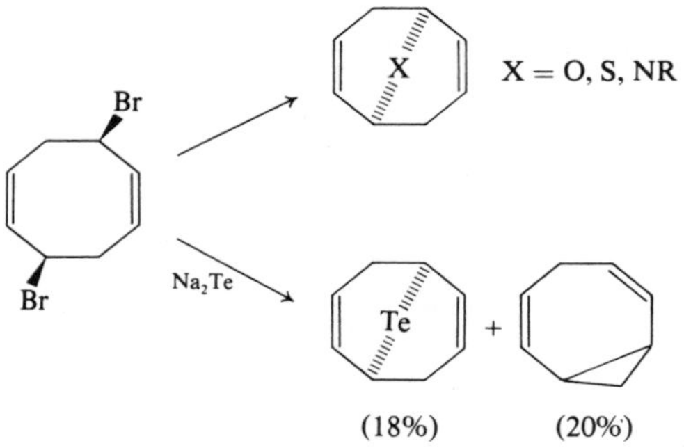

X = O, S, NR

Na₂Te

(18%)   (20%)

The popular Wittig reaction is a condensative elimination which is especially valuable for the preparation of (Z)-olefins under salt-free conditions (20). The contrathermodynamic stereoselectivity of this process has been pointed out by Schlosser in a lecture on α,α'-diastereogenic reactions (21). The term denotes that two chiral centers are created at the termini of the new bond which is to be formed.

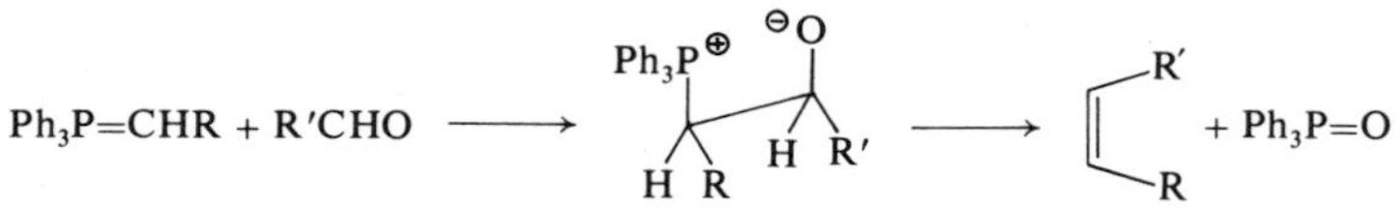

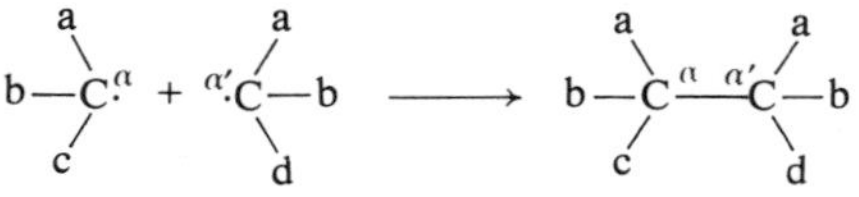

$$Ph_3P=CHR + R'CHO \longrightarrow \quad \longrightarrow \quad + Ph_3P=O$$

erythro + threo

The α,α'-diastereogenic reactions are often characterized by an energy profile in which the thermodynamically controlled pathway has a higher activation energy, and the kinetic product is formed via a lower energy transition state. The less stable product is often sterically encumbered. London forces have been invoked to explain the anomalous results, however, the total effect is not adequately accounted for by these forces alone. It may be that the composite characteristics of chemical softness are responsible for such dramatic manifestations. Spatial interactions among groupings of similar nature appear to have considerable influence on the energy barriers of the transition states.*

It should be mentioned that the keto phosphine oxide (1) adopts a more congested conformation (23) in order to allow pairwise spatial interactions be-

---

* For pertinent theoretical treatments of steric and nonbonded attractions, see Hoffmann *et al.* and Epiotis (22).

tween hydrogen and oxygen atoms four bonds apart. The hydrogen atoms are hardened by the neighboring electronegative groups.

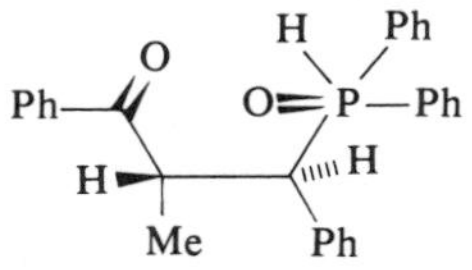

(1)

5.2. ADDITION TO DOUBLE BONDS

5.2.1. Homopolar Additions

Olefins are essentially soft donors. π Complexation of alkenes with soft metal ions, such as $Ag^{\oplus}$, $Pt^{4\oplus}$, and $Pd^{2\oplus}$, is well established. The addition of halogens and pseudohalogens is commonly regarded as proceeding via the π complexes, as detected spectroscopically for the olefin–bromine charge-transfer species (24).

Alkenes may undergo uncatalyzed dimerization which, according to the Woodward–Hoffmann rules, is nonconcerted. The salient feature is that the major products are often contrathermodynamic, e.g.,

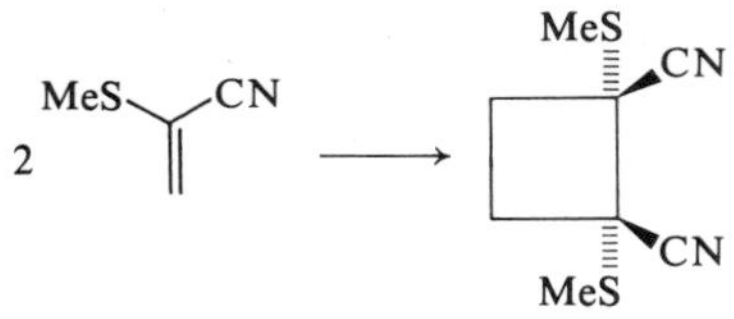

(Ref. 25)

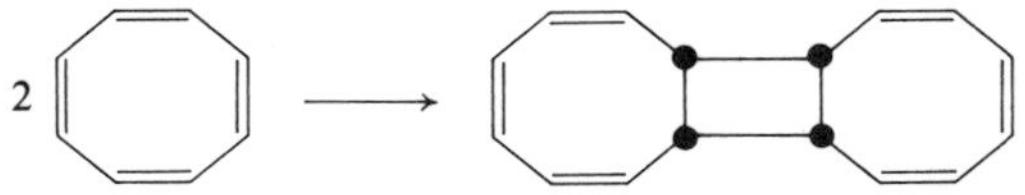

(Ref. 26)

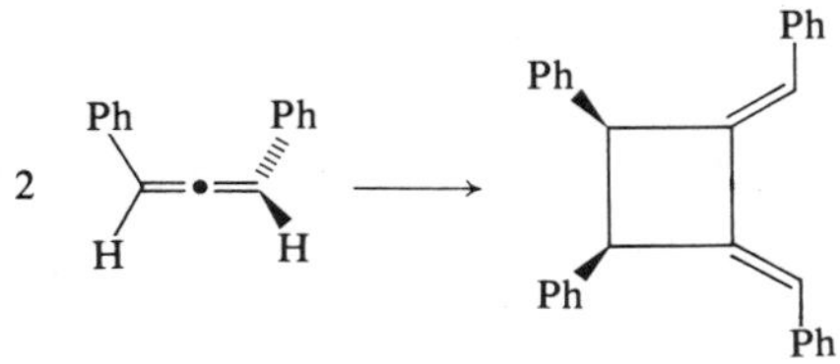

(Ref. 27)

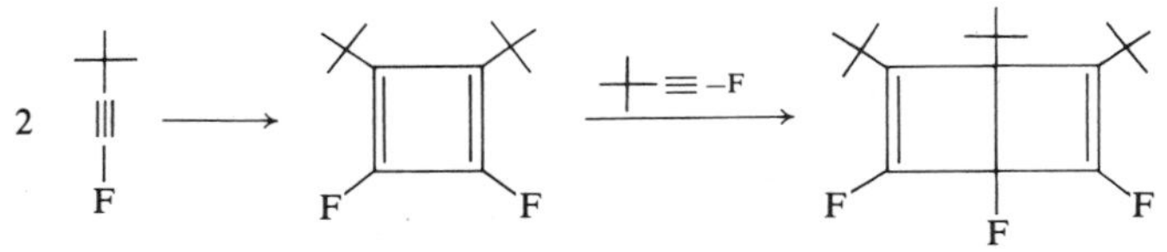

(Ref. 28)

As orbital control is implicit in the soft–soft interactions of the HSAB principle, it should be possible to define qualitatively the various aspects of the Diels–Alder reaction. Regiospecificity between dissymmetrical addends is found to arise from preferential formation of the first bond between the softest centers of the partners (29).

Product stereochemistry of [4+2]-cycloaddition, e.g., cyclopentadiene plus maleic anhydride, infers a lineup of the diacyloxy moiety of maleic anhydride with the central atoms of the diene. Thus, it is evident that the harder atoms lie in proximity in space and so do the soft components. Although secondary π-orbital interactions (30) are invoked to account for endo stereospecificity, these are not indispensable conditions. Cases which still obey the endo rule are known where direct π interactions are absent.

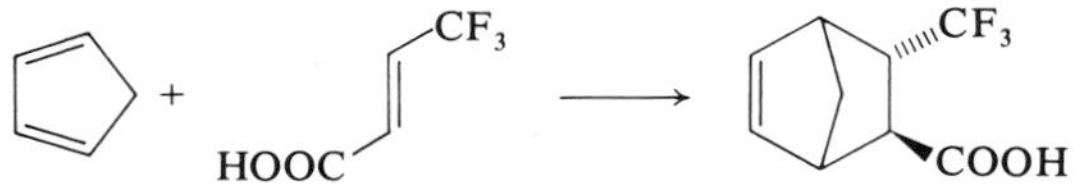

It should be emphasized that the present argument is not to discredit the secondary interactions when the requisite conditions are met, for the π interaction is actually inclusive in the general statement.

Selective 1,3-dipolar additions leading to the more crowded stereoisomers are also known, for example (31):

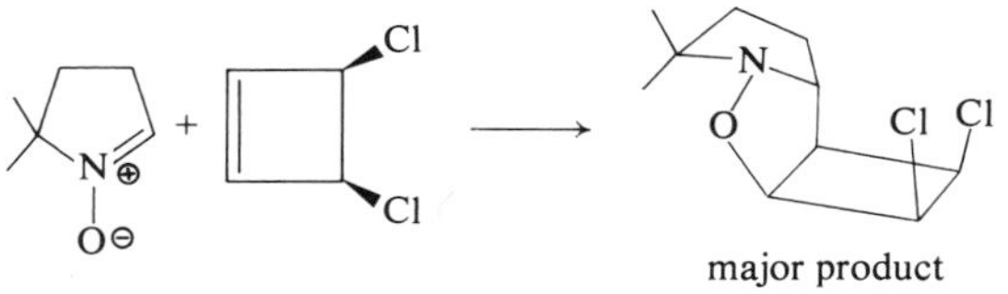

major product

Carbene addition to olefinic linkages to form cyclopropanes also shows stereoselectivity (32). The cis compounds are the principal or exclusive adducts when X = alkyl, aryl, halo, alkylthio, or alkylseleno group, while the trans adducts are formed preferentially when X = trimethylsilyl, alkoxy-carbonyl, or aryloxy. It is clear that carbenes containing a soft substituent X prefer the reaction path leading to products in which X and alkyl R are closer, and the opposite is true when X is hard.

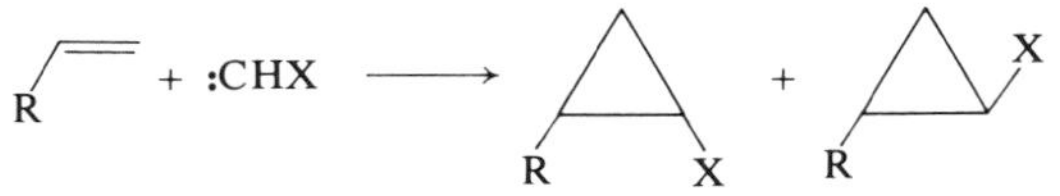

5.2.2. Heteropolar Additions

The ionic addition of unsymmetrical addends to alkenes obeys the Markovnikov rule. For the addition of HX, the Markovnikov transition state of C–H bond formation is encouraged by symbiosis. The creation of a more highly substituted carbocation is favored because the central carbon carries fewer soft hydrogen atoms. The carbenium ion may also be considered as an acid–base complex of a carbene–cation in which the carbene is a donor. Thus, the secondary carbenium ion $R_2CH^\oplus$ is made up of a resonance hybrid $R_2C:{\to}H^\oplus \leftrightarrow HRC:{\to}R^\oplus \leftrightarrow R^\oplus{\leftarrow}:CHR$, and the primary ion $RCH_2^\oplus$ is represented by the hybrid of $HRC:{\to}H^\oplus \leftrightarrow H^\oplus{\leftarrow}:CHR \leftrightarrow H_2C:{\to}R^\oplus$. The major difference in the two arises from the relative viability of $HRC:{\to}$ $R^\oplus$ vs. $HRC:{\to}H^\oplus$. The former is the more stable because $R^\oplus$ is softer than $H^\oplus$ and interacts better with the soft carbene.

Methoxybromination is a competing process during the addition of bromine to alkenes in methanol. Dubois *et al.* (33) argued that bromonium ions with highly substituted carbon atoms capable of carrying a larger share of positive charge are harder. The harder acceptors are, of course, more receptive to hard base (MeOH) than to the soft bromide ion.

Addition of thiocyanic acid to alkenes is rather interesting in view of the ambident nature of the thiocyanate ion. Norbornene gives a mixture containing 90% isothiocyanate and 9% thiocyanate (34), suggesting considerable hardness of the cationic carbon. In contrast, addition of HNCS to acrylic esters leads to thiocyanate adducts exclusively. The S end is employed to bond with the β carbon of the unsaturated acyl system. It should be noted that neutralization of tropylium fluoroborate with NaSCN yields cycloheptatrienyl isothiocyanate (35).

An ionic route is followed during addition of thiocyanogen to olefins. It seems likely that cyanoepisulfonium ions are the intermediates (36).

Dialkyl(alkylthio)sulfonium salts add across carbon–carbon double and triple bonds in a trans-stereospecific manner (37). Episulfonium salts are conveniently prepared by treatment of olefins with methyl(bismethylthio)sulfonium hexachloroantimonate (38). The displacement by olefin must occur at the softer sulfur atom, although it is impossible to ascertain whether the double bond attacks the sulfonium center in the reaction described by Helmkamp *et al.* (37).

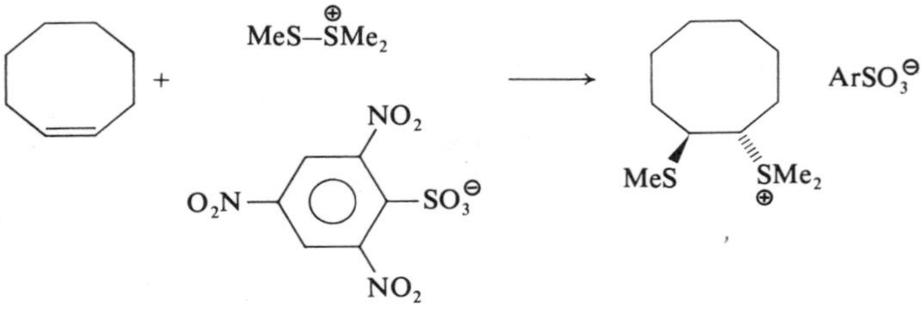

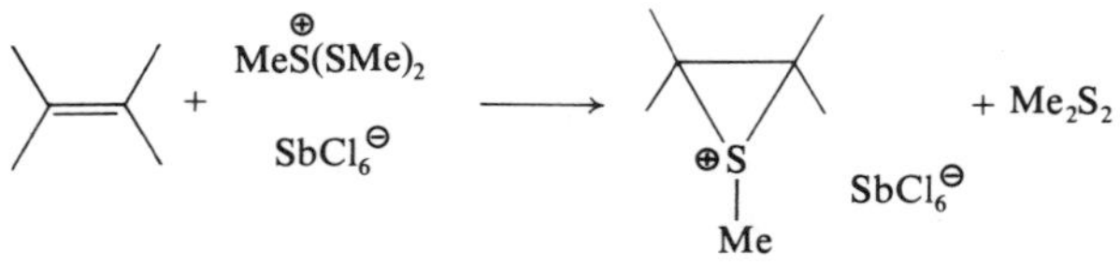

The addition of bromine to cyclooctatetraene at $-55°$ generates a stable, nonclassical homotropylium ion which is subsequently captured by $Br^{\ominus}$ stereospecifically (39). The seemingly disconcerting fact is that both bromine atoms approach the olefin from the more hindered face yielding the less stable product in an α,α'-diastereogenic reaction.

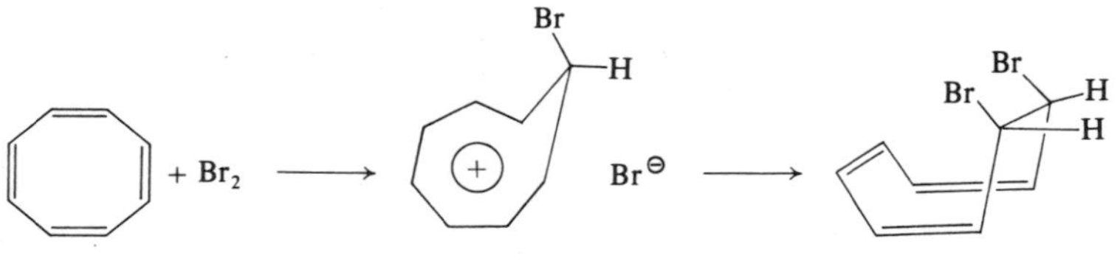

Solvatomercuration (40) is a polar addition which follows the Markovnikov rule. Because acetoxymercuric ion is a soft acid, the reaction proceeds facilely. Thallium ions are also soft, hence they attack multiple bonds efficiently. As Tl(III) is also an oxidant, the initial solvatothallation products often undergo rearrangements (41, 42) unobserved during solvatomercuration.

5.3. OXIDATION

Epoxidation of olefins with peracids (43) is of considerable synthetic value. The intrinsic instability of peracids caused by the soft–hard combination spurs the transfer of an active oxygen to soft bases.

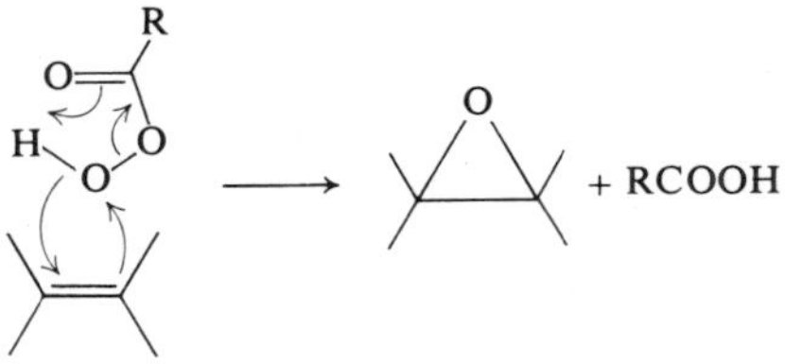

The metal ion-catalyzed decomposition of alkyl hydroperoxides in the presence of olefins also leads to the formation of oxiranes. The effectiveness of Mo(VI), Ti(IV), and W(VI) complexes is related to their being hard Lewis acids (44, 45).

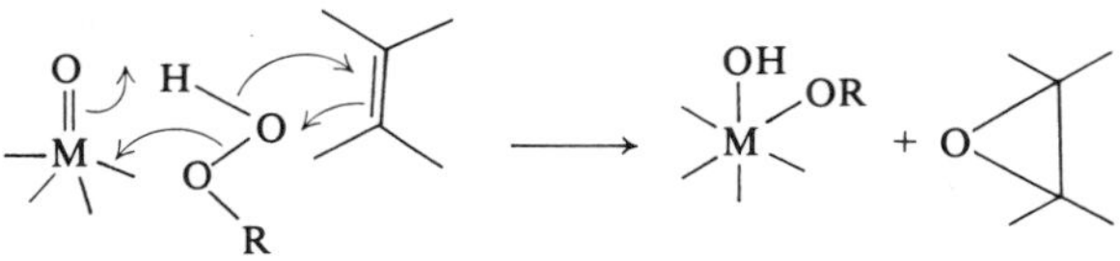

The synthesis of (Z)-1,2-diols by reaction of alkenes with permanganate (46, 47) and with osmium tetroxide is common practice. It is thought that a nucleophilic attack on the soft oxygen atom of the reagents by the π bond is involved in an essentially concerted cyclic process (48).

An unusual endo-hydroxylation of hexamethyl(Dewar)benzene (49) by OsO$_4$ has been reported. Perhaps coordination of the second double bond with the Os nucleus stabilizes the transition state.

Ruthenium tetroxide is an even more powerful oxidant (50, 51) for carbon–carbon double-bond cleavage. The escalated reactivity and the lessened stability of RuO$_4$ are the results of the softer metal atom being surrounded by hard oxygen atoms.

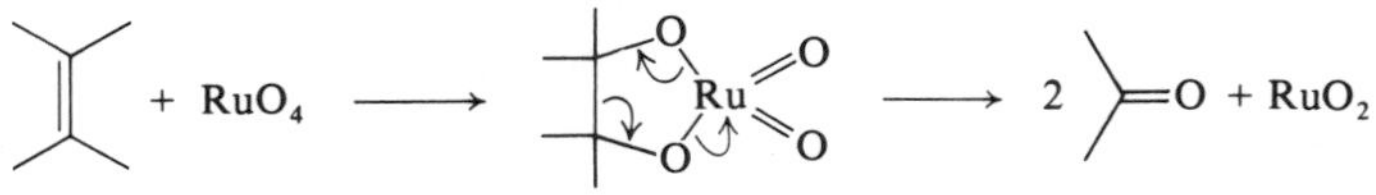

Cyclohexyl metaborate activates alkyl peroxides so that the combination provides effective epoxidizing agents (52). A good arrangement of soft–hard fragments ready for transfer of an oxygen to olefins is shown. The harder phenyl metaborate reacts with the hydroxy oxygen of tetralin hydroperoxide (53).

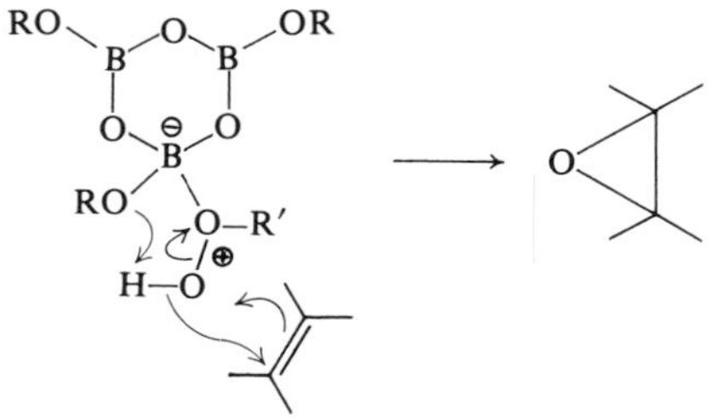

Tetracyanoethylene oxide is a potent enophile which often behaves as a 1,3-dipolar species. It is remarkable that benzene is attacked resulting in a cyclohexadiene derivative (54). The normal course can be diverted to a competing O-transfer process, however, when the olefin's reaction partners are very soft (55).

5.4. MISCELLANEOUS REACTIONS

The oxidation of δ,ϵ-unsaturated Bunte salts by iodine leads to tetrahydrothiophenes (56) predominantly; π participation is clearly indicated. The sulfur atom of the sulfenyl iodide intermediates is a very soft acceptor.

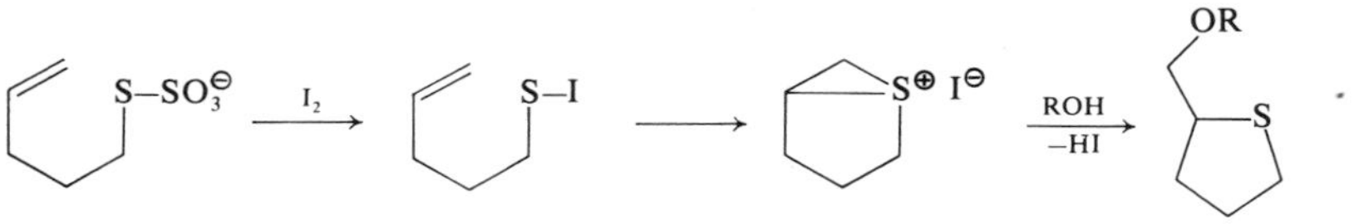

The addition of Grignard reagents to vinylsilanes (57) is generally favored by the softer R (tertiary > secondary > primary) in RMgX. Harder reagents such as benzyl-, phenyl-, and allylmagnesium halides effect only displacement at Si, provided that a leaving group is present.

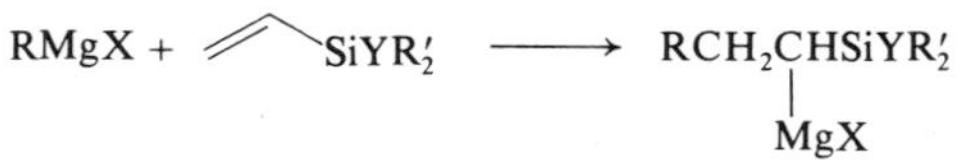

$$RMgX + \diagup\!\!\!\diagdown_{SiYR'_2} \longrightarrow RCH_2CHSiYR'_2$$
$$\underset{\overset{|}{MgX}}{}$$

The ease of carbanion formation from haloforms (58) depends on the α halogen as I > Br > Cl > F. This same order is valid for haloethylenes (59) as evidenced by the rate of deuterium exchange.

$$Cl_2C{=}C\overset{H}{\underset{Br}{\diagdown}} \; > \; Cl_2C{=}C\overset{H}{\underset{Cl}{\diagdown}} \; > \; Cl_2C{=}C\overset{H}{\underset{F}{\diagdown}}$$

The condensation of sodiostannanes with bromoacetylene gives bromoethy-nyltin derivatives (60), instead of the terminal acetylene. Apparently vinyl carbanion intermediates, specifically those stabilized by α-Br, are involved.

$$R_3SnNa + BrC{\equiv}CH \longrightarrow BrC{\equiv}CSnR_3$$

The chlorine atom of 2-phenyl-2-o-nitrophenyl-1-chloroethene can be replaced by nucleophiles (61). It reacts 10^9–10^{10} times faster with p-toluene-thiolate than with chloride ion.

The vinylic fluorine is specifically displaced by the hard methoxide ion from a chlorofluorocyclopentene derivative (62). The existence of transition state symbiosis,

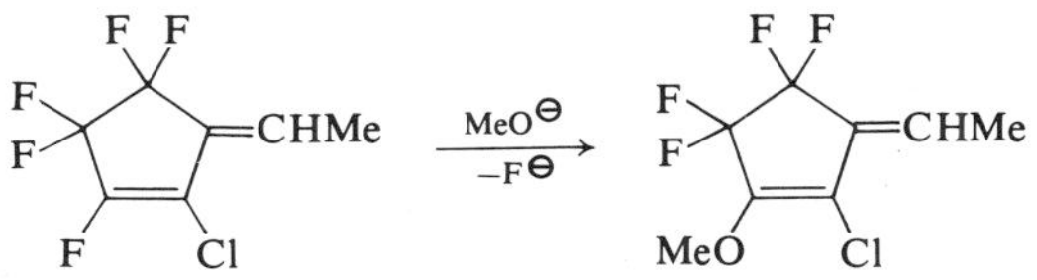

as well as the fact that chlorine stabilizes the carbanion better, must be the con-tributing factors to the specificity.

The S_N2' displacement of fluoropropenes by halide ions $Y^{\ominus}$ (63), as shown below, is especially favored by fluoride ion ($F^{\ominus} \gg Cl^{\ominus} > Br^{\ominus}, I^{\ominus}$). Symbiosis could be the reason.

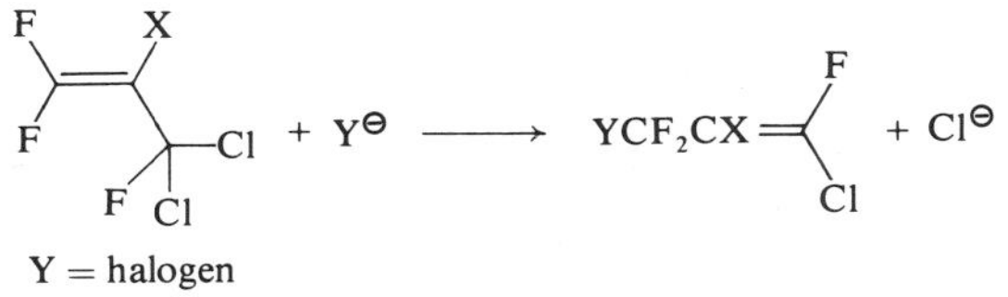

Y = halogen
X = H, Cl

Treatment of 3,3-difluorotetrachloropropene with $AlBr_3$ gives, among other products, 1-bromo-1,2-dichloro-3,3,3-trifluoropropene (64). The proposed mechanism invokes successive specific attack of soft ($Br^{\ominus}$) and hard ($F^{\ominus}$) ions at the softer and harder carbocation centers, respectively. The source of $F^{\ominus}$ is from other concurrent processes.

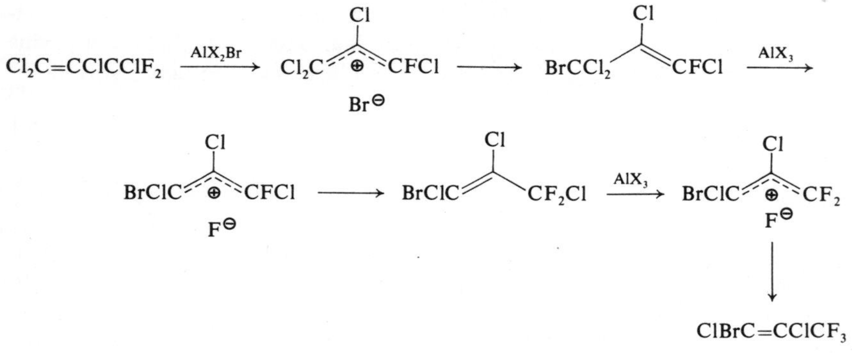

$$Cl_2C=CClCClF_2 \xrightarrow{AlX_2Br} Cl_2C\overset{Cl}{\underset{\overset{|}{Br^{\ominus}}}{\overset{\oplus}{{=}{=}}}}CFCl \longrightarrow BrCCl_2\overset{Cl}{\underset{}{\overset{|}{=}}}CFCl \xrightarrow{AlX_3}$$

$$BrClC\overset{Cl}{\underset{\overset{|}{F^{\ominus}}}{\overset{\oplus}{{=}{=}}}}CFCl \longrightarrow BrClC\overset{Cl}{\underset{}{\overset{|}{=}}}CF_2Cl \xrightarrow{AlX_3} BrClC\overset{Cl}{\underset{\overset{|}{F^{\ominus}}}{\overset{\oplus}{{=}{=}}}}CF_2$$

$$ClBrC=CClCF_3$$

Allylic rearrangement of 3-chloropentafluoropropene goes through an unsymmetrical cation, in contrast to the expectation of ionizing the softer $Cl^{\ominus}$. Symbiosis indicates, however, that formation of $SbF_4Cl_2^{\ominus}$ is better than $SbF_3Cl_3^{\ominus}$.

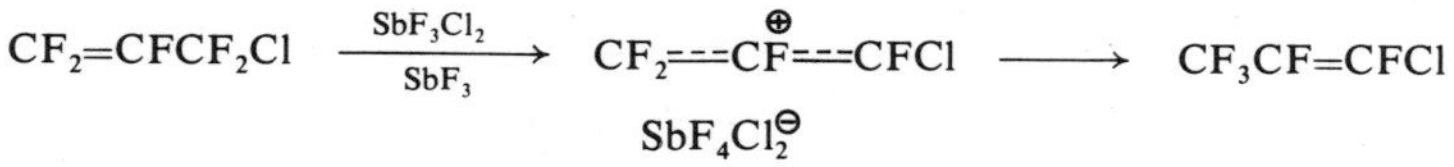

$$CF_2=CFCF_2Cl \xrightarrow[\text{SbF}_3]{\text{SbF}_3Cl_2} CF_2\overset{\oplus}{{=}{=}{=}}CF{=}{=}{=}CFCl \longrightarrow CF_3CF=CFCl$$
$$SbF_4Cl_2^{\ominus}$$

The catalytic effect of olefins on chlorination of nitroalkanes was discovered during an attempt to add tert-butyl hypochlorite across the styrene double bond (65). It appears that a four-centered reaction is facilitated by the soft π complexation as outlined below.

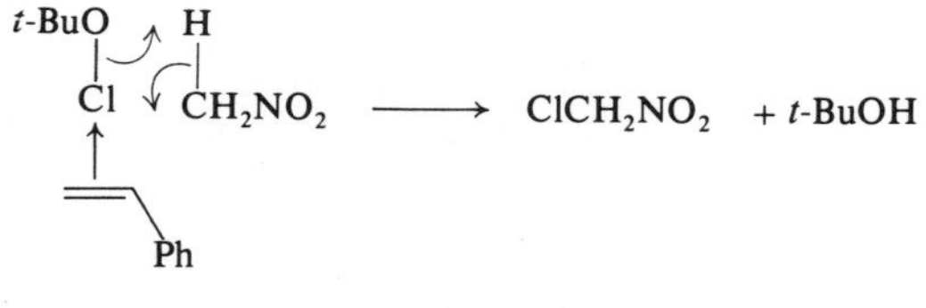

$$\begin{array}{c} t\text{-BuO} \quad\curvearrowright\, H \\ \;\;|\qquad\;\;| \\ Cl \;\;\vee\; CH_2NO_2 \longrightarrow ClCH_2NO_2 \;+ t\text{-BuOH} \\ \uparrow \\ \overset{\shortparallel}{\underset{Ph}{}} \end{array}$$

Although the ene reaction (66) is a concerted homopolar process, it is favored by the presence of polar groups in the enophiles. For example, formaldehyde is an active enophile, however, it hardly participates in the Diels–Alder reaction.

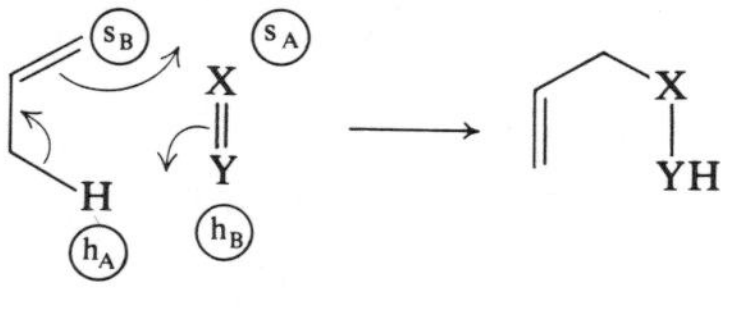

Strikingly, hexafluoroacetone undergoes an ene reaction with propene (67) yielding a sulfide instead of a thiol. The thionic sulfur atom acts as a soft acceptor as predicted by the HSAB principle.

The substituent effects on [1,5]sigmatropic ester shift (68) have been scrutinized. The k_{rel} increases with the hardness of the substituent R geminal to the

E = COOMe

migrating ester group. Perhaps the transition state of the rearrangement resembles the product in which R becomes attached to a harder sp^2-hybridized carbon atom.

REFERENCES

1. W. H. Saunders, Jr. and A. F. Cockerill, "Mechanisms of Elimination Reactions." Wiley, New York, 1973.
2. C. H. DePuy and R. W. King, Chem. Rev. 60, 431 (1960).
3. A. C. Cope and E. R. Trumbull, Org. React. 11, 317 (1960).
4. B. Ganem, J. Org. Chem. 40, 1998 (1975).
5. C. A. Kingsbury and D. J. Cram, J. Am. Chem. Soc. 82, 1810 (1960).
6. B. M. Trost and T. N. Salzmann, J. Am. Chem. Soc. 95, 6840 (1973).
7. J. Hayami, N. Ono, and A. Kagi, Tetrahedron Lett. p. 1385 (1968).
8. F. Naso and L. Ronzini, J. Chem. Soc., Perkin Trans. 1, p. 340 (1974).
9. R. G. Pearson, J. Am. Chem. Soc. 71, 2212 (1949).
10. D. A. Evans, C. A. Bryan, and G. M. Wahl, J. Org. Chem. 35, 4122 (1970).
11. E. J. Corey and S. Terashima, Tetrahedron Lett. p. 111 (1972).
12. R. A. Bartsch, K. E. Wiegers, and D. M. Guritz, J. Am. Chem. Soc. 96, 430 (1974).
13. J. F. King and R. G. Pews, Can. J. Chem. 42, 1294 (1964).
14. S. Winstein, D. Pressman, and W. G. Young, J. Am. Chem. Soc. 61, 1645 (1939).
15. I. J. Borowitz, D. Weiss, and R. K. Crouch, J. Org. Chem. 36, 2377 (1971).
16. W. K. Kwok and S. I. Miller, J. Org. Chem. 35, 4034 (1970).
17. F. G. Bordwell and B. B. Jarvis, J. Am. Chem. Soc. 95, 3585 (1973).
18. E. Cuthbertson and D. D. MacNicol, J. Chem. Soc., Perkin Trans. 1, p. 1893 (1974).
19. E. Cuthbertson and D. D. MacNicol, Chem. Commun. p. 498 (1974).
20. M. Schlosser, G. Muller, and K. F. Christmann, Angew. Chem., Int. Ed. Engl. 5, 667 (1966).
21. M. Schlosser, Bull. Soc. Chim. Fr. p. 453 (1971).

22. R. Hoffmann, C. C. Levin, and R. A. Moss, *J. Am. Chem. Soc.* **95**, 629 (1973); N. D. Epiotis, *J. Am. Chem. Soc.* **95**, 3087 (1973).
23. R. O. Day, V. W. Day, and C. A. Kingsbury, *Tetrahedron Lett.* p. 3041 (1976).
24. J. E. Dubois and F. Garnier, *Chem. Commun.* p. 241 (1968).
25. K.-D. Gundermann and R. Huchting, *Chem. Ber.* **92**, 415 (1959).
26. G. Schröder and W. Martin, *Angew. Chem., Int. Ed. Engl.* **5**, 130 (1966).
27. E. V. Dehmlow, *Chem. Ber.* **100**, 3260 (1967).
28. H. G. Viehe, R. Merényi, J. F. M. Oth, and P. Valange, *Angew. Chem., Int. Ed. Engl.* **3**, 746 (1964).
29. O. Eisenstein, J. M. Lefour, and Nguyen Trong Anh, *Chem. Commun.* p. 969 (1971).
30. R. Hoffmann and R. B. Woodward, *J. Am. Chem. Soc.* **87**, 4388 (1965).
31. C. DeMicheli, A. Gamba-Invernizzi, and R. Gandolfi, *Tetrahedron Lett.* p. 2493 (1975).
32. R. A. Moss, *in* "Selective Organic Transformations" (B. S. Thyagarayan, ed.), Vol. 1, pp. 35–88. Wiley (Interscience), New York, 1970.
33. J. E. Dubois, M. H. Durand, G. Mouvier, and J. Chrétien, *Tetrahedron Lett.* p. 2993 (1975).
34. L. A. Spurlock and P. E. Newallis, *J. Org. Chem.* **30**, 2086 (1965).
35. H. Kessler and A. Walter, *Angew. Chem., Int. Ed. Engl.* **12**, 773 (1973).
36. R. G. Guy, R. Bonnett, and D. Lanigan, *Chem. Ind. (London)* p. 1702 (1969).
37. G. K. Helmkamp, B. A. Olsen, and J. R. Koskinen, *J. Org. Chem.* **30**, 1623 (1965).
38. G. Capozzi, O. DeLucchi, V. Lucchini, and G. Modena, *Tetrahedron Lett.* p. 2603 (1975).
39. R. Huisgen and G. Boche, *Tetrahedron Lett.* p. 1769 (1965).
40. H. C. Brown and P. J. Geoghegan, *J. Org. Chem.* **35**, 1844 (1970).
41. A. McKillop, J. D. Hunt, E. C. Taylor, and F. Kienzle, *Tetrahedron Lett.* p. 5275 (1970).
42. A. McKillop, O. H. Oldenziel, B. P. Swann, E. C. Taylor, and R. L. Robey, *J. Am. Chem. Soc.* **93**, 7331 (1971).
43. D. Swern, *Org. React.* **7**, 378 (1953).
44. R. A. Sheldon and J. A. VanDoorn, *J. Catal.* **31**, 427 (1973).
45. K. B. Sharpless, J. M. Townsend, and D. R. Williams, *J. Am. Chem. Soc.* **94**, 295 (1972).
46. D. G. Lee and J. R. Brownridge, *J. Am. Chem. Soc.* **95**, 3033 (1973).
47. K. W. Wiberg, C. J. Deutsch, and J. Rocek, *J. Am. Chem. Soc.* **95**, 3034 (1973).
48. J. S. Littler, *Tetrahedron* **27**, 81 (1971).
49. G. R. Krow and J. Reilly, *Tetrahedron Lett.* p. 3129 (1972).
50. D. G. Lee and M. van den Engh, *in* "Oxidation in Organic Chemistry" (W. S. Trahanovsky, ed.), Part B, Chapter 4. Academic Press, New York, 1973.
51. J. E. Lyons and J. O. Turner, *J. Org. Chem.* **37**, 2881 (1972).
52. P. F. Wolf and R. K. Barnes, *J. Org. Chem.* **34**, 3441 (1969).
53. P. F. Wolf, J. E. McKeon, and D. W. Cannell, *J. Org. Chem.* **40**, 1875 (1975).
54. W. J. Linn and R. E. Benson, *J. Am. Chem. Soc.* **87**, 3657 (1965).
55. P. Brown and R. C. Cookson, *Tetrahedron* **24**, 2551 (1968).
56. R. D. Rieke, S. E. Bales, and L. C. Roberts, *Chem. Commun.* p. 974 (1972).
57. G. R. Buell, R. Corriu, C. Guerin, and L. Spialter, *J. Am. Chem. Soc.* **92**, 7424 (1970).
58. J. Hine, N. W. Burske, M. Hine, and P. B. Langford, *J. Am. Chem. Soc.* **79**, 1406 (1957).
59. D. Daloze, H. G. Viehe, and G. Chiurdoglu, *Tetrahedron Lett.* p. 3925 (1969).
60. S. V. Zavgorodnii and A. A. Petrov, *Zh. Obshch. Khim.* **35**, 931 (1965).
61. P. Beltrame, P. L. Beltrame, G. Carboni, and M. L. Cereda, *J. Chem. Soc. B* p. 730 (1970).
62. J. D. Park and R. J. McMurtry, *Tetrahedron Lett.* p. 1301 (1971).
63. J. H. Fried and W. T. Miller, Jr., *J. Am. Chem. Soc.* **81**, 2078 (1959).
64. D. J. Burton and G. C. Briney, *J. Org. Chem.* **35**, 3036 (1970).
65. V. L. Heasley, G. E. Heasley, M. R. McConnell, K. A. Martin, D. M. Ingle, and P. D. Davis, *Tetrahedron Lett.* p. 4819 (1971).

66. H. M. R. Hoffmann, *Angew. Chem., Int. Ed. Engl.* **8**, 556 (1969).
67. W. J. Middleton and W. H. Sharkey, *J. Org. Chem.* **30**, 1384 (1965).
68. J. Backes, R. W. Hoffmann, and F. W. Steuber, *Angew. Chem., Int. Ed. Engl.* **14**, 553 (1975).

6

Aromatic and Heterocyclic Chemistry

6.1. EXCHANGE REACTIONS

Aryl halides, except fluorides, undergo metal–halogen exchange (1) with rates proportional to the polarizability of the C–X bond. The fact that p-cyanophenyllithium is obtained from treatment of the corresponding bromoarene with butyllithium (2) attests to the preference for soft interaction.

The conversion of bromobenzene to benzene-d_1 by successive treatment with n-BuLi amd D_2O is impractical. However, the reaction is improved by adding the organometallic reagent to the substrate in D_2O-saturated ether (3). Apparently the metal–halogen exchange proceeds much faster than the destruction of the reagent by D_2O. The presence of Me_3SiCl instead of heavy water in the medium allows for the preparation of trimethylsilylbenzene in good yield. The key to success for these procedures lies in the fact that both D_2O and Me_3SiCl are hard acids, whereas organometallic compounds are soft bases. Butyllithium attacks bromobenzene selectively to generate a harder base which reacts with the electrophile at a more leisurely pace.

The Ullmann biphenyl synthesis (4, 5) involves arylcopper intermediates. The coupling is a typical soft–soft interaction. A useful extension to the generation of alkylarenes is by the reaction of dialkylcuprates with haloarenes (6).

Metal–metal exchange (7) between organometallic compounds may be effected in the presence of bromo substituents on the partners containing very soft metal atoms, e.g.,

$$(p\text{-BrC}_6\text{H}_4)_2\text{Hg} + 2n\text{-BuLi} \longrightarrow 2p\text{-BrC}_6\text{H}_4\text{Li} + n\text{-Bu}_2\text{Hg}$$

6.2. ELECTROPHILIC SUBSTITUTIONS

Electrophilic aromatic substitution ($S_E Ar$) is the major mode of reaction for aromatic compounds as a result of the presence of a π-electron cloud which is a good donor. When very hard substituents are present, the nucleophilicity of the π system is reduced. Pentafluorotoluene represents an extreme case which undergoes electrophilic substitution at the methyl group (8).

The Friedel–Crafts alkylation of aromatics with ω-haloalkyl fluorides always favors the formation of ω-haloalkylarenes (9, 10). Symbiosis determines the selective C–X bond severance. Similarly, the indene synthesis by alkylation of benzene with 1-bromo-1-fluorocyclopropanes is shown to proceed via an electrocyclic ionization of the hard $F^{\ominus}$ (11). This behavior is opposite to the trend of the uncatalyzed thermal heterolysis of halocyclopropanes (see Chapter 11, Section 11.4).

$$ArH + F(CH_2)_n X \xrightarrow{BF_3} Ar(CH_2)_n X$$

$$X = Cl, Br, I$$

Fluorination of aromatics is conveniently achieved by using fluoroxy compounds, e.g., $FOCF_3$ (12, 13). The fluorine in these reagents is a very soft acceptor.

Iodination of phenol (14) involves an iodocyclohexadienone intermediate. The reversibility ratio for this process is larger than that for the corresponding reaction with 4-nitrophenol. The nitro group evidently makes the corresponding dienone intermediate harder so that loss of a proton to the water solvent occurs more easily than the removal of $I^{\oplus}$ by $I^{\ominus}$ in the former case.

Aromatic iodination via thallation has wide applicability (15). The iodine atom enters the same position as the thallium. Replacement of the hard trifluoroacetoxy groups of the thallium derivatives with iodide is favored because Tl(III) is soft.

The relative rates of $S_E Ar$ reactions of alkylbenzenes are opposite to those expected from the electron release effect of the alkyl groups (Baker–Nathan effect) (16). The rationalization of this phenomenon in terms of the HSAB principle is as follows. The stability of the σ-complex intermediate may be analyzed by considering the R–C$^\oplus$ fragment of the cross-conjugated canonical form. This fragment, as a carbene–carbenium ion pair [C: → R$^\oplus$], is more stable if R$^\oplus$ is softer, i.e., Me$^\oplus$ > Et$^\oplus$ > i-Pr$^\oplus$ > t-Bu$^\oplus$, in agreement with the Baker–Nathan effect.

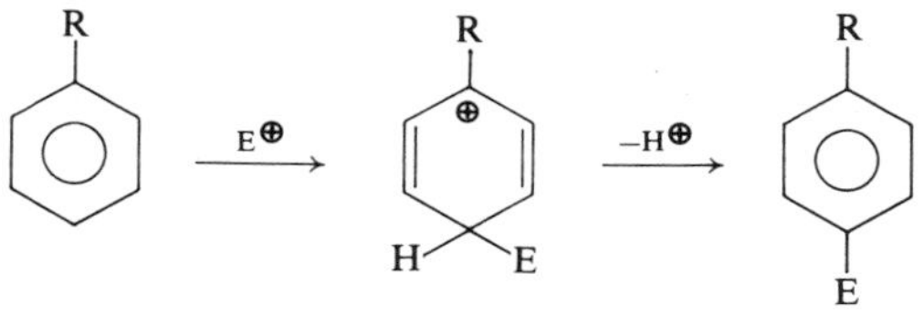

A similar reasoning may be applied to explain the benzene ring activation toward deuteration by the (dicarbonyl-π-cyclopentadienyliron)methyl group (17).

The stereochemical aspect of electrophilic substitution of metallocenes is intriguing. In acylation, the attack comes from the outside (anti to the metal) and the order of reactivities is ferrocene > ruthenocene > osmocene (18). On the other hand, during mercuration the electrophile approaches from the inside (syn to the metal). It is clear that (i) hard electrophiles align themselves farthest from the essentially soft metal in the transition states, and the charge-controlled reactions (acylation) occur more readily with the harder metallocenes, and (ii) soft electrophiles (e.g., Hg$^{2\oplus}$) approach the aromatic ring from the same side as the metal which permits soft–soft interaction.

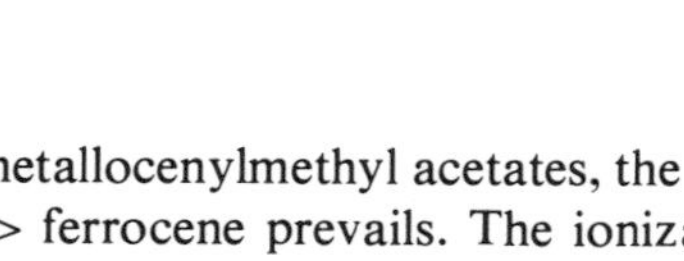

In the solvolysis of α-metallocenylmethyl acetates, the reactivity gradient osmocene > ruthenocene > ferrocene prevails. The ionization step may be envisioned as involving assistance from an electron pair of the ring. The geometry and electronic state of the metallocene moiety in the transition state must therefore resemble that in a substitution reaction by a soft acid.

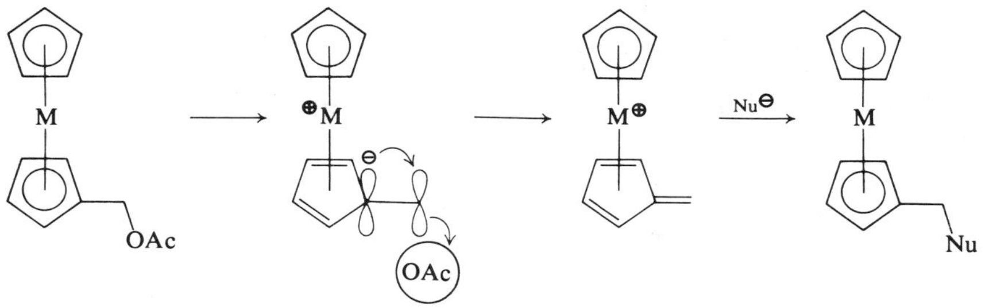

The oxygenation of aromatic compounds by Lewis acid-catalyzed decomposition of peroxydicarbonates (19–21) may be formulated as indicated in the following reaction. Saville's rule is applicable to these reactions.

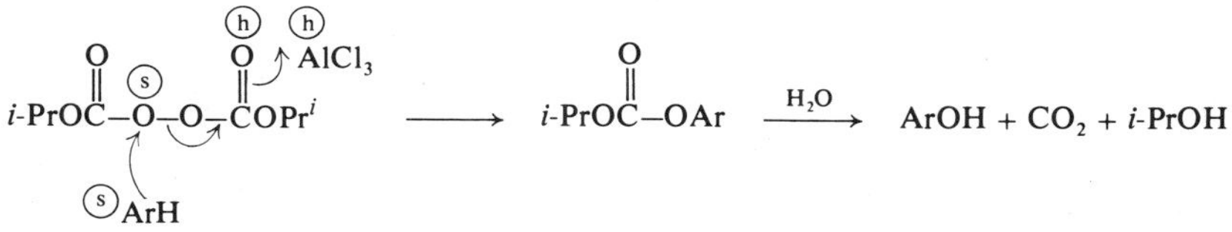

Although the mechanism of the Dakin reaction given in many textbooks does not involve participation of the π electron of the aromatic ring, the mandatory presence of polar substituents (e.g., OH) in the ortho or para position to the formyl group suggests that an internal electrophilic substitution of the adduct intermediate is a better representation. The uncatalyzed S_EAr process is a soft–soft interaction.

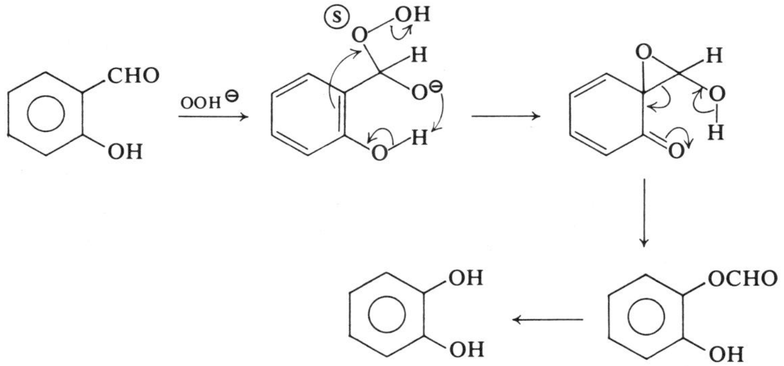

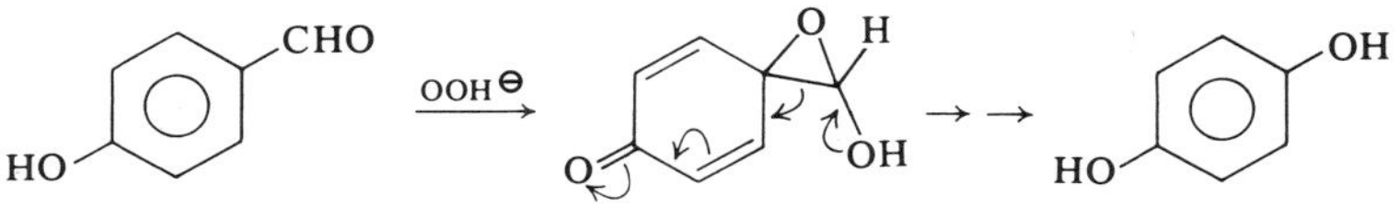

Diaryliodonium halides undergo first-order decomposition to give aryl iodide and aryl halides (22). The influence of the counterion, $I^\ominus > Br^\ominus > Cl^\ominus$, on rate is consistent with symbiotic stabilization of the transition state. The relatively soft acceptor role of the ipso carbon atom is also indicated.

6.3. NUCLEOPHILIC SUBSTITUTIONS

Generally nucleophilic aromatic substitution (S_NAr) takes place only when strongly electron-withdrawing groups are present. Recently it has been discovered that π-complexed metal carbonyls also render the aromatic ring electrophilic. Surprisingly the rate-determining step in aminodefluorination (23) of (fluorobenzene) chromium tricarbonyl is the loss of $F^\ominus$. The intermediate must be stabilized by symbiosis at two different sites: the ipso carbon which carries two hard substituents, and the chromium atom which now acquires electron donation from the more effective negative charge.

Semmelhack *et al.* (24) found that the halo leaving group is unnecessary, and a hydride ion can be formally displaced from the benzene ring of the chromium tricarbonyl complex. Evidence seems to be in favor of initial attack on the chromium atom which is also a soft–soft interaction.

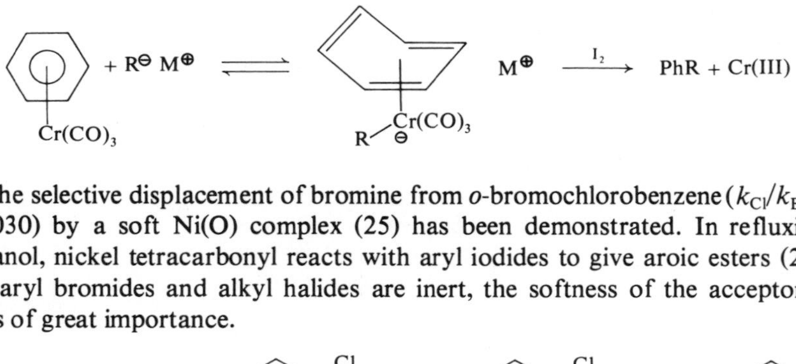

The selective displacement of bromine from *o*-bromochlorobenzene ($k_{Cl}/k_{Br} = 0.0030$) by a soft Ni(O) complex (25) has been demonstrated. In refluxing ethanol, nickel tetracarbonyl reacts with aryl iodides to give aroic esters (26). As aryl bromides and alkyl halides are inert, the softness of the acceptor is thus of great importance.

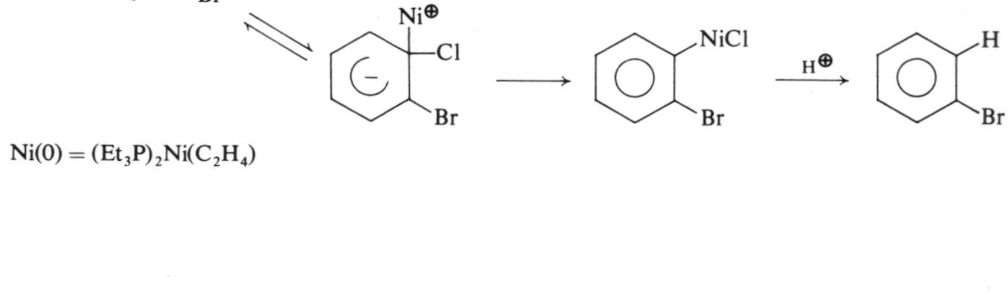

$Ni(0) = (Et_3P)_2Ni(C_2H_4)$

The mobility pattern for halonitrobenzenes (27) is $F \gg Cl > Br > I$ when they are displaced with nucleophilic atoms of the first row elements; the order is reversed when heavy atom bases are employed. On the basis of the HSAB principle, initial attack by hard bases is favored when the leaving halogen is hard because the first transition state is symbiotically stabilized. The opposite is true for soft nucleophiles.

The expulsion of halogen from halonitrobenzenes by the nitrite ion (28) always furnishes phenolic products. It has been shown by isotope labeling techniques that nitrogen attack is faster than oxygen attack when the leaving group is Cl, Br, or I, but the initial products are converted into the nitrite esters and then to the phenolates. Oxygen attack is faster if fluorine is being displaced. The results clearly show the existence of transition state symbiosis.

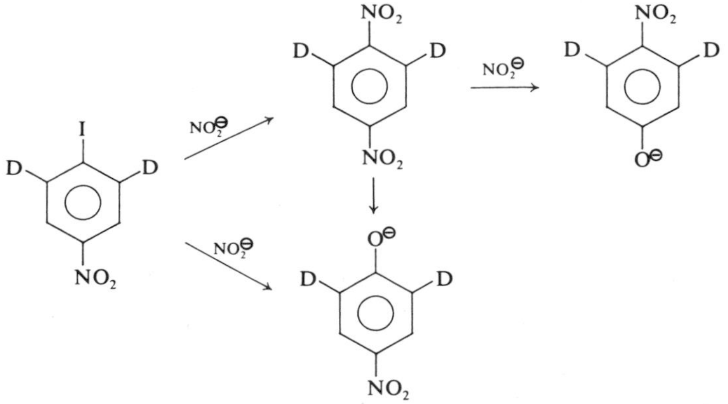

In the $S_N Ar$ reaction of polynitrophenyl derivatives (ArX) with the thiocyanate ion, the nature of the leaving group X has profound influence on the reaction rates (29). The rate ratio k_S/k_N has a 10^5 range (from X = pyridinium ion to X = I), with the iodo compound enormously favoring the attack by the soft S end of thiocyanate. It was suggested that transition state symbiosis is much more reasonable in $S_N Ar$ reactions because the entering and leaving groups are bonded to a tetrahedral carbon and are much closer to each other than they are in the aliphatic $S_N 2$ transition state.

The $S_N Ar$ reaction of polyhalo aromatic compounds has been discussed (30). For the displacement of fluorine para to substituent R, the relative stability of the transition state appears to follow the order $H \sim I > Br > Cl > F$, in contrast to the normal electron-attracting influence exerted by the halogen. The stability sequence is directly related to the chemical softness of the substituents. The fragment $R–C^{\ominus}$ may be regarded as a soft–soft combination $[R^{\ominus} \rightarrow \overset{..}{C}]$.

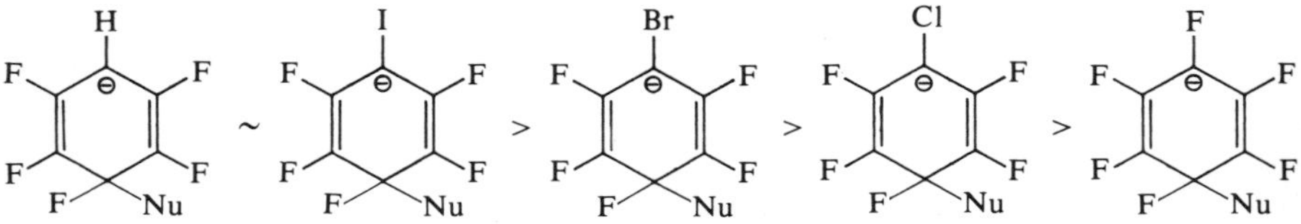

Variation of the substitution site according to the nature of the attacking base is known (31). The dichotomy may be the result of symbiotic stabilization of the transition state.

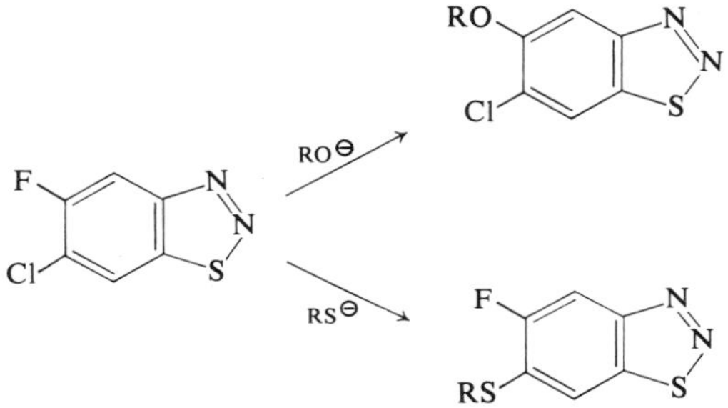

In nucleophilic substitution of 2,4-dinitro-1-thiocyanatobenzene, the mobility of the thiocyano group is similar to iodide. On the other hand, the fast reaction with methoxide ion is attributed to the attack at the harder cyano carbon instead of the aromatic ring (32).

Despite their high electrophilicity, arynes are discriminatory toward nucleophiles (33). An interesting reaction involving an additive rearrangement of benzyne with alkylidenetriphenylphosphoranes (34) is depicted as the following.

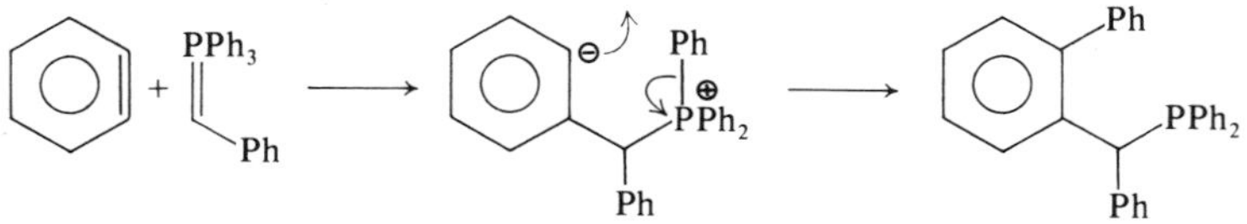

Benzyne displays greatly different reactivities toward $I^\ominus$, $Br^\ominus$, $Cl^\ominus$, and ethanol (35). The strong hard bases (e.g., $NH_2^\ominus$) cannot compete with the thiolate anions for benzyne (36), and the trapping efficiency for 9,10-dehydrophenanthrene (37) is a function of donor softness: $PhSLi > PhLi > C_5H_{10}NLi > Ph(Me)NLi > PhOLi$.

3,4-Dehydropyridine (pyridyne) generated by dehydrohalogenation of 3- or 4-halopyridine with hard bases in the presence of methanethiol (38) gives equimolar amounts of 3- and 4-methylthiopyridines predominantly.

Abramovitch and Newman (39) found that the displacement rate ratios k_{MeS}/k_{MeO} of 0.71 for 2-fluoropyridine and 4.7 for 2-bromopyridine agree with the HSAB principle. The F/Br mobility ratios were also deduced for the methoxide ion (28.5) and for the methanethiolate (4.3). The diminished E_{act} for the bromo compound is due to attractive dispersion forces between the polarizable nucleophile and the leaving group (40).

2-Bromo-5-nitro-1,3,4-thiadiazole yields two different "monosubstituted" products from its reactions with sodium and silver benzenethiolates, respectively (41). The counterion in the nucleophile dictates the site of attack by the sulfide anion through the selective coordination with the leaving group more compatible in softness. Thus, Ⓢ:Ⓢ silver–bromine and Ⓗ:Ⓗ sodium–oxygen (of the nitro group) pairings are responsible for the outcome.

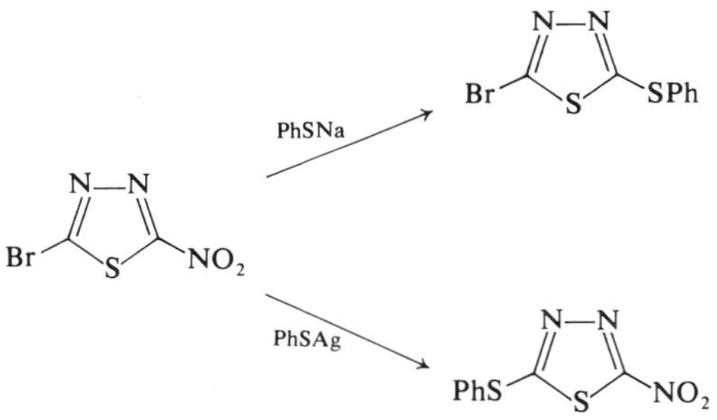

2-Bromo- and 2-iodoimidazoles are reductively dehalogenated by the sulfite ion. The corresponding chloro and fluoro compounds are stable to this soft reagent (42). It should be noted that the fluorine is easily replaced by hard bases via an S_NAr mechanism.

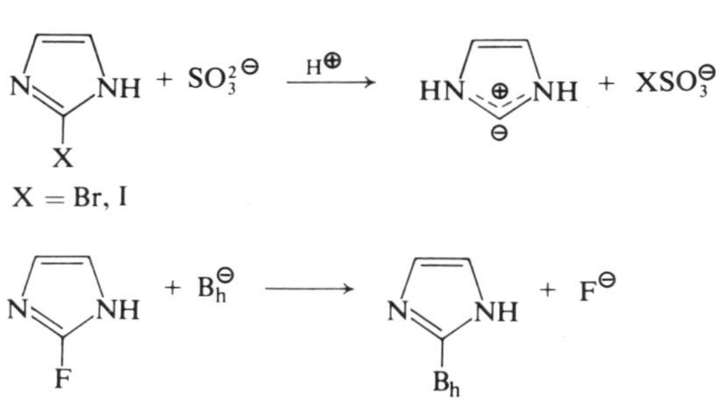

The hard nucleophiles ($F^\ominus$, $RO^\ominus$) displace the $MeSO_2$ groups in 3,5-dichloro-2,6-bis(methylsulfonyl)pyridine, whereas the softer $CN^\ominus$ and R_2NH species displace the chlorine atoms (43).

Dihydropyridine derivatives are obtained by the addition of nucleophiles to pyridinium salts (44). In general, the harder bases add to C-2, whereas the softer bases enter C-4, of the heterocyclic system (45–47).

Dithionite ion is a very soft Lewis base and it transfers a sulfoxylate ion to the C-4 of the pyridinium salts (48).

The reaction of N-phenacylpyridinium bromide with lithium dialkylcuprates (49) results in the isolation of acetophenone. The fate of the heterocyclic moiety has apparently not been determined. It is reasonable to assume, on the basis of the soft nature of these organometallic reagents, that 4-alkylpyridines must have been produced concomitantly. Addition of chloroformate esters to a mixture of pyridine and a cuprate reagent leads predominantly to N-carbo-alkoxy-4-alkyl(aryl)-1,4-dihydropyridines (50).

5-Nitroisoquinoline methiodide gives a covalent pseudobase immedi-

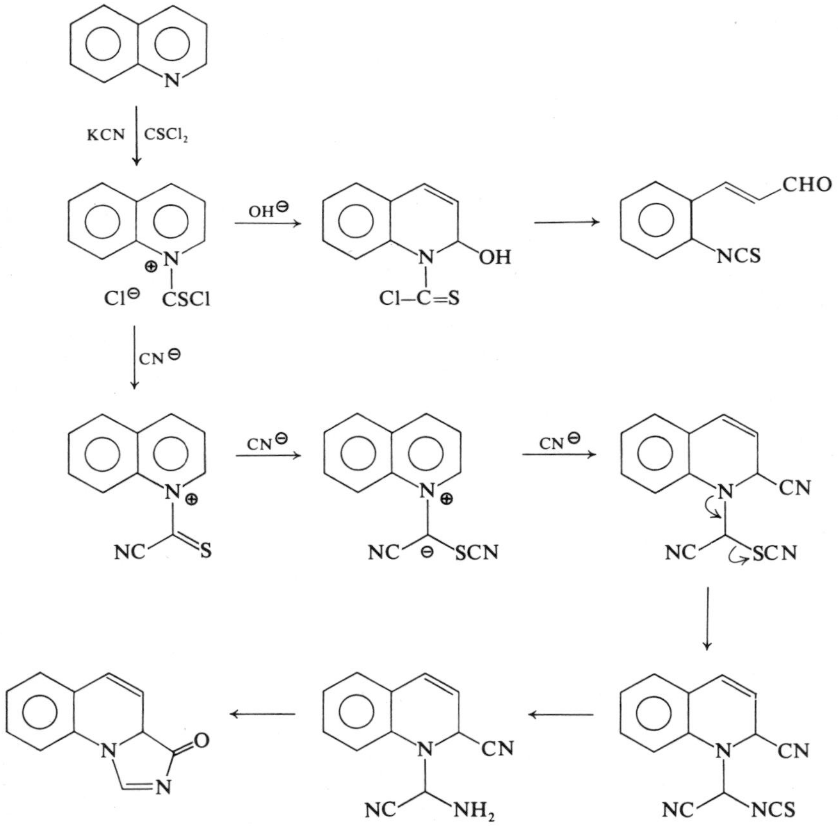

SCHEME 6.1

ately on treatment with silver oxide (51), whereas ordinary *N*-methyliso-
quinolinium salts afford the strongly alkaline quaternary hydroxides which are
slowly transformed into the covalent form. The nitro group at C-5 clearly
enhances the hardness of C-1 of the heterocycle, thereby favoring the attack
by hard bases such as H_2O and $HO^{\ominus}$. The nitroisoquinoline does not yield a
Reissert compound (52), for the softer cyanide ion is superseded by $HO^{\ominus}$ or
H_2O.

Quinoline is converted to the Reissert compound. However, sodium benzoate is also produced. In the competing process quinoline is released from the
acyl appendage of the intermediate by the hard base.

When quinoline is reacted with a combination of thiophosgene and KCN, *o*-
isothiocyanato-(*E*)-cinnamaldehyde and 3-oxoimidazo[1,5-*a*]quinoline are obtained (53) according to Scheme 6.1. It should be noted that the chlorothio-
carbonyl function, which is quite soft, reacts readily with cyanide ion in
preference to hydroxide.

The general validity of the HSAB principle is again borne out by the reaction
pattern of *N*-alkoxypyridinium salts (54). The α hydrogen from the methylene
group adjoining the oxygen atom is removed by hard alkoxide ions. On the
other hand, soft anions ($I^{\ominus}$, $SCN^{\ominus}$, $S_2O_3^{2\ominus}$) displace pyridine oxide [$S_N2(C)$],
and cyanide ion adds to C-2, leading in the formation of 2-cyanopyridine.

N-(*p*-Nitrophenoxy) pyridinium salts aryloxylate methoxybenzenes and an-
nelate with benzonitrile (55). The oxygen site of the electrophiles is a softer acceptor than the ortho carbon, because as acids the core atoms belonging to the
higher main groups are often softer (e.g., $O^{\oplus} > N^{\oplus} > C^{\oplus}$). It should be
emphasized that for the same group the lighter electrophile atom would be softer, e.g., $>C{=}O > >C{=}S$ (56).

Metals supply a very soft electron to the C-4 of pyridines or pyridinium ions (57). Interestingly, viologens can be reduced by cyanohydrin anions (58).

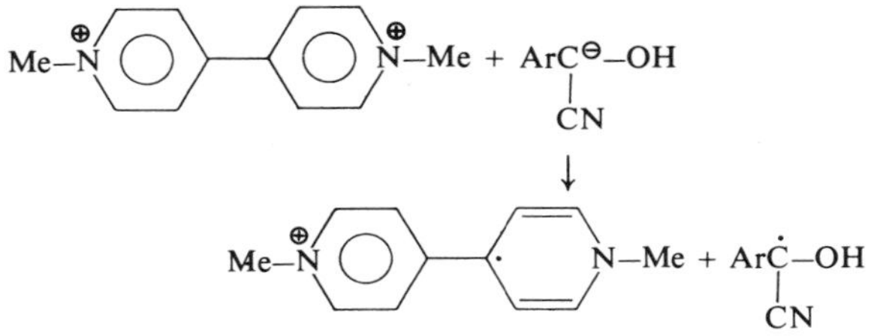

The Birch reduction also involves electron transfer to aromatic rings. In order to understand the substituent effect on the sites of reduction, it is instructive to consider the relative stabilities of the following structures:

The disubstituted carbanion favored for the hydride–carbene complex is more stable than the carbanion–carbene complex corresponding to the tertiary carbanion.

The same influence on orientation is, however, not exerted by every type of substituent. For example, benzoic acid is reduced to 1,4-dihydrobenzoic acid, presumably owing to an initial electron transfer to the carbonyl oxygen.

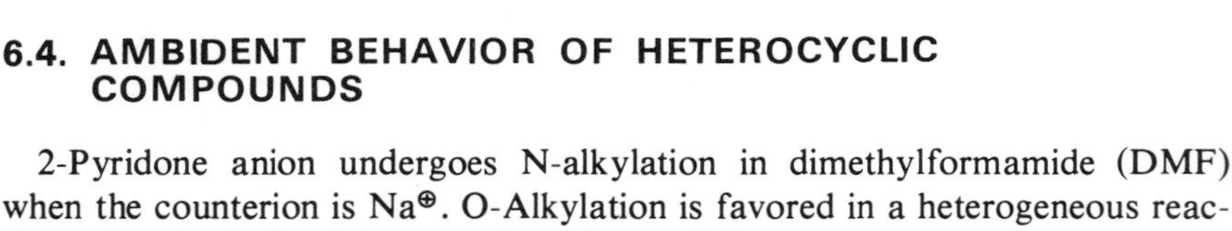

6.4. AMBIDENT BEHAVIOR OF HETEROCYCLIC COMPOUNDS

2-Pyridone anion undergoes N-alkylation in dimethylformamide (DMF) when the counterion is Na$^\oplus$. O-Alkylation is favored in a heterogeneous reaction using the Ag$^\oplus$ salt in nonpolar solvents (59).

The alkylation of 4-hydroxypyrimidines (60) has also been studied. The results are readily explicable by the HSAB principle. Acetylation of the thallium(I) salt of 2-pyridone (61) at −40° provides up to 40% of the *N*-acetyl derivative, which on warming rearranges largely to the O-isomer. The Na salt of phenanthridinone reacts with benzoyl chloride at room temperature to afford only the less stable O-benzoylated product (62).

Thus, the use of the soft Tl(I) salt diverts a portion of acetylation to the nitrogen site, even though the process is charge-controlled. In contrast, the hard Na salts of pyridone derivatives favor O-acylation.

1-(Carboxymethoxy)-4-pyridone decomposes on treatment with diazomethane to yield 4-methoxypyridine, formaldehyde, and carbon dioxide (63). Both carboxylate ion and the 4-pyridone oxygen atom are hard, thus the observation is not in conflict with the HSAB concept.

The reactions of 2- and 4-aminopyridines with ambident electrophiles, namely, alkoxycarbonyl isothiocyanates, have been analyzed within the HSAB context (64). It appears that the field effect also plays an important role.

6.5. AROMATIC CLAISEN REARRANGEMENTS

In the aromatic Claisen rearrangement (65), the ether–phenol transformation occurs at about 200°; S–C allyl shifts are more difficult to achieve (66). As the formation of the dienone intermediates is rate-determining, the analysis of perturbance around the heteroatoms (X) leading to the intermediates should provide clues to the relative rates, for changes are the same elsewhere. Thus, the crux of the problem involves only the net change of a $C_{sp^3}-X$ to a $C_{sp^2}-X$ bond. Such a change is favored with X = O rather than that with X = S because trigonal carbon is harder.

Since amino-Claisen rearrangement is facilitated by Lewis acids (67), it is therefore conceivable that the sulfonium salts derived from allyl aryl sulfides would undergo rearrangement at lower temperatures. Such an effect due to prehardening of sulfur atom has been observed (68).

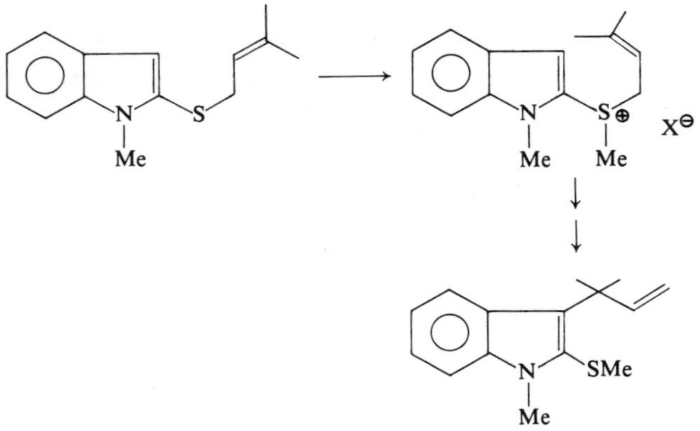

An interesting retro-thio-Claisen rearrangement (69) has been reported. The driving force is the inherent instability of the C=S bond of the thioamide.

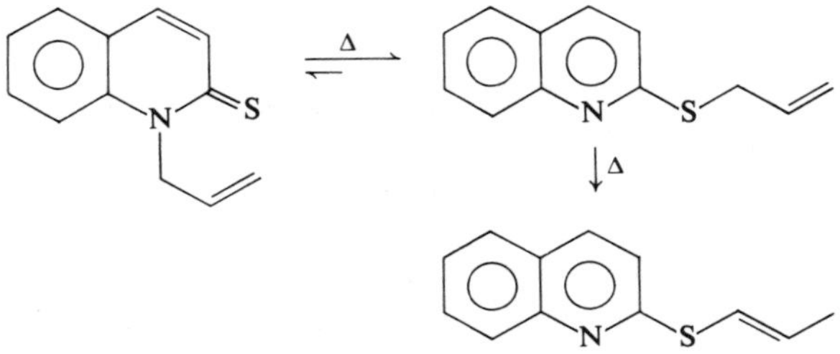

It is pertinent to note that the rearrangement of $Ar_2\overset{\ominus}{C}XCH_2CH=CH_2$ $Li^{\oplus}$ to $Ar_2C(XLi)CH_2CH=CH_2$ is spontaneous at low temperature when X = O, and it can be induced only by refluxing in tetrahydrofuran when X is the softer NMe group (70).

6.6. OXIDATIVE COUPLING OF PHENOLS

The oxidative coupling of phenols (71) is biogenetically significant. This reaction can be performed *in vitro* by oxidants such as $FeCl_3$ and $K_3Fe(CN)_6$. Although both reagents are one-electron oxidants, disparate results often occur.

The oxidation of 2-naphthol by ferricyanide occurs via an outer-sphere process in which no direct bonding between the substrate and the metal ion exists; the ensuing phenoxy radical is relatively free. On the other hand, the complex having an ArO—Fe bond appears to intervene in the ferric chloride reaction. As the phenoxy radical remains coordinated to Fe, coupling can occur only at the carbon.

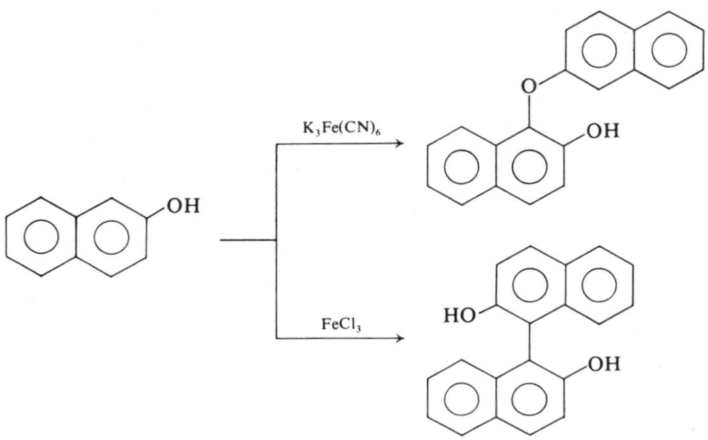

The bifurcation is of HSAB origin. The iron atom in ferricyanide is quite soft since it is surrounded by six cyanide ligands and formal negative charges are present in the anionic cluster. Consequently, the interaction of the iron with the hard oxygen of phenol (phenolate) is unfavorable (symbiotically destabilized). The iron atom of ferric chloride is hard, and the exchange of its ligands with phenols is easy.

REFERENCES

1. R. G. Jones and H. Gilman, *Org. React.* **6**, 339 (1951).
2. H. Gilman and D. S. Melstrom, *J. Am. Chem. Soc.* **70**, 4177 (1948).
3. R. Taylor, *Tetrahedron Lett.* p. 435 (1975).
4. F. Ullmann, *Justus Liebigs Ann. Chem.* **332**, 38 (1904).
5. P. E. Fanta. *Chem. Rev.* **38**, 39 (1946).
6. E. J. Corey and G. H. Posner, *J. Am. Chem. Soc.* **89**, 3911 (1967).
7. H. Gilman and R. G. Jones, *J. Am. Chem. Soc.* **63**, 1443 (1941).
8. G. A. Olah and Y. K. Mo, *J. Am. Chem. Soc.* **95**, 6827 (1973).
9. N. O. Calloway, *J. Am. Chem. Soc.* **59**, 1474 (1937).
10. G. A. Olah and S. J. Kuhn, *J. Org. Chem.* **29**, 2317 (1964).
11. C. Müller and P. Weyerstahl, *Tetrahedron* **31**, 1787 (1975).
12. D. H. R. Barton, A. Ganguly, R. H. Hesse, S. N. Loo, and M. M. Pechet, *Chem. Commun.* p. 806 (1968).
13. D. H. R. Barton, R. H. Hesse, H. T. Toh, and M. M. Pechet, *J. Org. Chem.* **37**, 329 (1972).
14. E. Grovenstein, Jr., N. S. Aprahamian, C. J. Bryan, N. S. Gnanapragasam, D. C. Kilby, J. M. McKelvey, Jr., and R. J. Sullivan, *J. Am. Chem. Soc.* **95**, 4261 (1973).
15. A. McKillop, J. D. Hunt, M. J. Zelesko, J. S. Fowler, E. C. Taylor, G. McGillivray, and F. Kienzle, *J. Am. Chem. Soc.* **93**, 4841 (1971).

16. J. W. Baker and W. S. Nathan, *J. Chem. Soc.* p. 1840 (1935).
17. S. N. Anderson, D. H. Ballard, and M. D. Johnson, *Chem. Commun.* p. 779 (1971).
18. J. A. Mangravite and T. G. Traylor, *Tetrahedron Lett.* p. 4461 (1967).
19. P. Kovacic and M. E. Kurz, *J. Am. Chem. Soc.* **87**, 4811 (1965).
20. P. Kovacic and M. E. Kurz, *Chem. Commun.* p. 321 (1966).
21. P. Kovacic and S. T. Morneweck, *J. Am. Chem. Soc.* **87**, 1566 (1965).
22. F. M. Beringer and M. Mausner, *J. Am. Chem. Soc.* **80**, 4535 (1958).
23. J. F. Bunnett and H. Hermann, *J. Org. Chem.* **36**, 4081 (1971).
24. M. F. Semmelhack, H. T. Hall, M. Yoshifuji, and G. Clark, *J. Am. Chem. Soc.* **97**, 1247 (1975).
25. D. R. Fahey, *J. Am. Chem. Soc.* **92**, 402 (1970).
26. N. L. Bauld, *Tetrahedron Lett.* p. 1841 (1963).
27. J. Miller, "Aromatic Nucleophilic Substitution," p. 140. Am. Elsevier, New York, 1968.
28. T. J. Broxton, D. M. Muir, and A. J. Parker, *J. Org. Chem.* **40**, 2037 (1975).
29. D. E. Giles and A. J. Parker, *Aust. J. Chem.* **26**, 273 (1973).
30. J. Burdon, *Tetrahedron* **21**, 3373 (1965).
31. J. H. Davies, E. Haddock, P. Kirby, and S. B. Webb, *J. Chem. Soc. C* p. 2843 (1971).
32. J. Miller and F. H. Kendall, *J. Chem. Soc., Perkin Trans. 2*, p. 1645 (1974).
33. R. W. Hoffmann, "Dehydrobenzene and Cycloalkynes." Academic Press, New York, 1967.
34. E. Zbiral, *Monatsh. Chem.* **95**, 1760 (1964).
35. G. Wittig and R. W. Hoffmann, *Chem. Ber.* **95**, 2729 (1962).
36. J. F. Bunnett and T. K. Brotherton, *J. Org. Chem.* **23**, 904 (1958).
37. R. Huisgen, *in* "Organometallic Chemistry" (H. Zeiss, ed.), p. 36. Van Nostrand-Rheinhold, Princeton, New Jersey, 1960.
38. J. A. Zoltewicz and C. Nisi, *J. Org. Chem.* **34**, 765 (1969).
39. R. A. Abramovitch and A. J. Newman, Jr., *J. Org. Chem.* **39**, 3692 (1974).
40. D. L. Dalrymple, J. D. Reinheimer, D. Barnes, and R. Baker, *J. Org. Chem.* **29**, 2647 (1964).
41. H. Newman, E. L. Evans, and R. B. Angier, *Tetrahedron Lett.* p. 5829 (1968).
42. K. L. Kirk, W. Nagai, and L. A. Cohen, *J. Am. Chem. Soc.* **95**, 8389 (1973).
43. T. J. Giacobbe and S. D. McGregor, *J. Org. Chem.* **39**, 1685 (1974).
44. U. Eisner and J. Kuthan, *Chem. Rev.* **72**, 1 (1972).
45. F. Kröhnke, K. Ellegast, and E. Bertram, *Justus Liebigs Ann. Chem.* **600**, 176 (1957).
46. W. v. E. Doering and W. E. McEwen, *J. Am. Chem. Soc.* **73**, 2104 (1951).
47. D. C. Dittmer and J. M. Kolyer, *J. Org. Chem.* **28**, 1720 (1963).
48. H. E. Dubb, M. Saunders, and J. H. Wang, *J. Am. Chem. Soc.* **80**, 1767 (1958).
49. G. H. Posner and J.-S. Ting, *Synth. Commun.* **4**, 355 (1974).
50. E. Piers and M. Soucy, *Can. J. Chem.* **52**, 3563 (1974).
51. A. Claus and K. Hoffmann, *J. Prakt. Chem.* [NS] **47**, 252 (1893).
52. B. C. Uff, J. R. Kershaw, and S. R. Chhabra, *J. Chem. Soc., Perkin Trans. 1*, p. 1146 (1974).
53. R. Hull, *J. Chem. Soc. C*, p. 1777 (1968).
54. A. R. Katritzky and E. Lunt, *Tetrahedron* **25**, 4291 (1969).
55. R. A. Abramovitch, M. Inabasekaran, and S. Kato, *J. Am. Chem. Soc.* **95**, 5428 (1973).
56. N. J. Turro and V. Ramamurthy, *Tetrahedron Lett.* p. 2423 (1976).
57. B. Emmert, *Ber. Dtsch. Chem. Ges.* **50**, 31 (1917); **52**, 1351 (1919).
58. D. N. Kramer, G. G. Guilbault, and F. M. Miller, *J. Org. Chem.* **32**, 1163 (1967).
59. G. C. Hopkins, J. P. Jonak, H. J. Minnemeyer, and H. Tieckelmann, *J. Org. Chem.* **32**, 4040 (1967).
60. J. P. Jonak, G. C. Hopkins, H. J. Minnemeyer, and H. Tieckelmann, *J. Org. Chem.* **35**, 2512 (1970).

61. A. McKillop, M. J. Zelesko, and E. C. Taylor, *Tetrahedron Lett.* p. 4945 (1968).
62. D. Y. Curtin and J. H. Engelmann, *Tetrahedron Lett.* p. 3911 (1968).
63. H. N. Bojarska-Dahlig and H. J. den Hertog, *Recl. Trav. Chim. Pays-Bas* **77**, 331 (1958).
64. T. Matsui and M. Nagano, *Chem. Pharm. Bull.* **22**, 2123 (1974).
65. A. Jefferson and F. Scheinmann, *Q. Rev., Chem. Soc.* **22**, 391 (1968).
66. H. Kwart and J. L. Schwartz, *Chem. Commun.* p. 44 (1969), and references cited therein.
67. C. D. Hurd and W. D. Jenkins, *J. Org. Chem.* **22**, 1418 (1957).
68. B. W. Bycroft and W. Landon, *Chem. Commun.* p. 967 (1970).
69. Y. Makisumi and T. Sasatani, *Tetrahedron Lett.* p. 1975 (1969).
70. M. T. Reetz and D. Schinzer, *Tetrahedron Lett.* p. 3485 (1975).
71. W. I. Taylor and A. R. Battersby, eds., "Oxidative Coupling of Phenols." Dekker, New York, 1967.

7

Reactivity of Carbonyl Compounds

7.1. THE CARBONYL GROUP AS A HARD ACCEPTOR AND A HARD DONOR

The carbonyl group has a hard donor oxygen and a fairly hard acceptor carbon. For electrophilic attack, hard bases are preferred. The ordinary carbonyl derivatives such as hydrazones, oximes, and semicarbazones are the condensation products derived from hard–hard interactions. Soft bases such as alkylphosphines do not attack the carbonyl carbon under normal conditions.

3-Tosyloxypivaldehyde has two electrophilic centers of different hardness. It is noteworthy that the soft benzenethiolate displaces the tosyloxy group, whereas the harder cyanide ion adds to the carbonyl (1). The tosyloxy group is then extruded from the cyanohydrin in a symbiotically favored process.

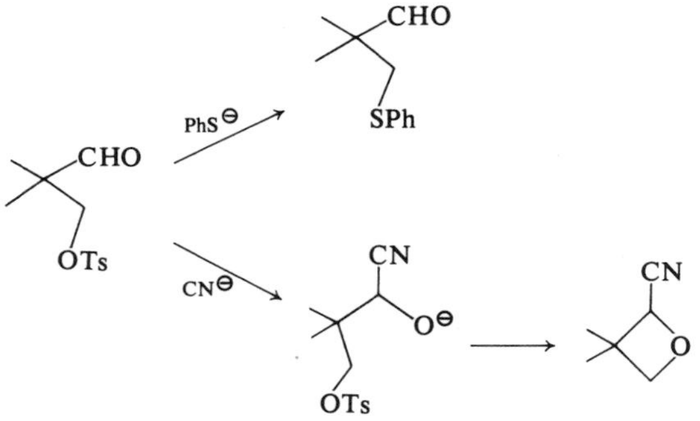

7.1.1. Baeyer–Villiger Oxidation

The Baeyer–Villiger oxidation (2) of aldehydes with peracids almost always yields carboxylic acids indicating a specific hydride shift during the collapse of the tetrahedral adducts. The soft hydride is a better group for migration to the soft oxygen acceptor site. For unsymmetrical ketones, the approximate migratory aptitude for the alkyl groups follows the softness order tertiary > secondary > primary. It is interesting to note that nortricyclanone gives the lactone exclusively from the migration of the cyclopentane group (3). Of course, the migrating C–C bond is softer than the alternative one since the exocyclic orbitals of cyclopropane are known to possess a high s character and the bonds formed from these orbitals tend to be less polarizable and harder.

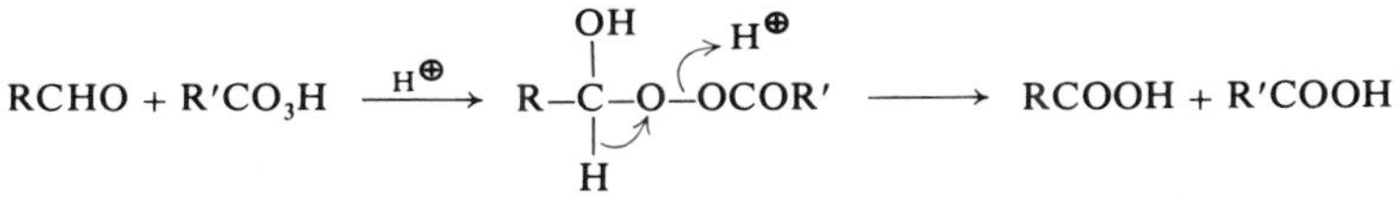

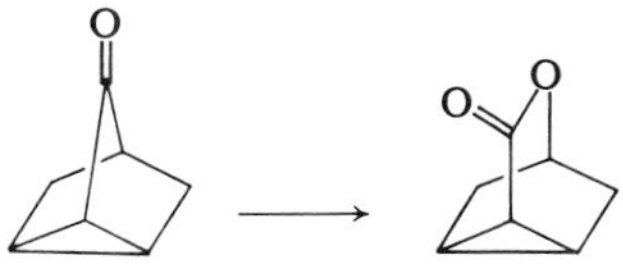

$$RCHO + R'CO_3H \xrightarrow{H^\oplus} R-\underset{\underset{H}{|}}{\overset{\overset{OH}{|}}{C}}-O-OCOR' \longrightarrow RCOOH + R'COOH$$

The greater migratory aptitude of aryl groups over the alkyls during the Baeyer–Villiger oxidation and related processes is due to the participation of the soft π system.

The oxidation of α-diketones to anhydrides by peracids appears to involve the hard carbonyl as migrant. However, a mechanism obviating the inconsistency may be formulated as follows.

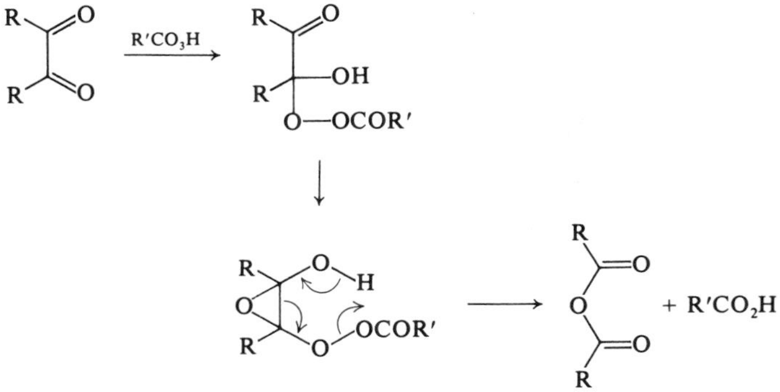

Aldehydes are also oxidized to the corresponding carboxylic acids by either alkaline hydrogen peroxide or silver oxide. Saville's rule 1 is obeyed in both cases.

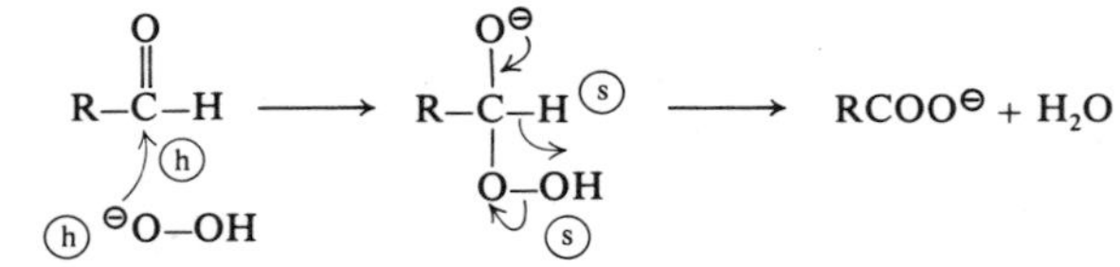

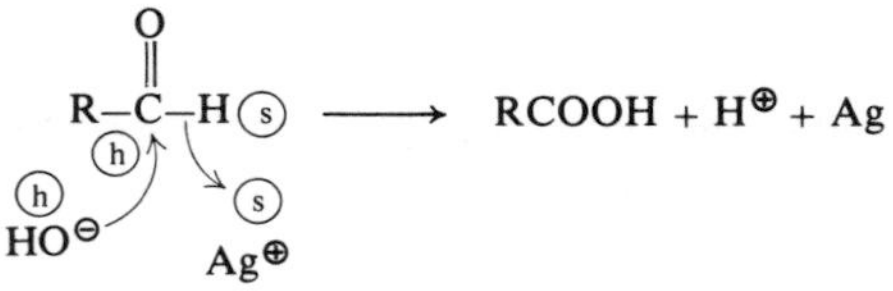

$$R-\overset{\overset{\displaystyle O}{\|}}{C}-H \longrightarrow R-\overset{\overset{\displaystyle O^{\ominus}}{|}}{\underset{\underset{\displaystyle O-OH}{|}}{C}}-H\ \text{(s)} \longrightarrow RCOO^{\ominus} + H_2O$$

$$R-\overset{\overset{\displaystyle O}{\|}}{C}-H\ \text{(s)} \longrightarrow RCOOH + H^{\oplus} + Ag$$

7.1.2. Favorskii Rearrangement

The conversion of α-haloketones to esters [Favorskii rearrangement (4)] by alkoxide ions usually occurs via the Loftfield mechanism (5).

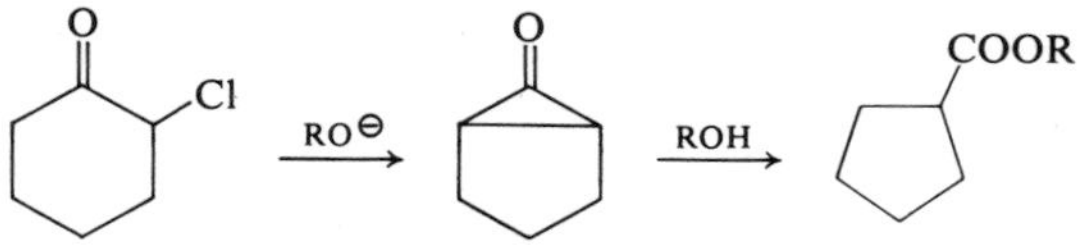

Treatment of α,α'-dibromoketones with lithium diorganocuprates gives rise to the Loftfield intermediates which undergo ring opening to generate enolate ions (6). Every step of the sequence represents a soft–soft interaction.

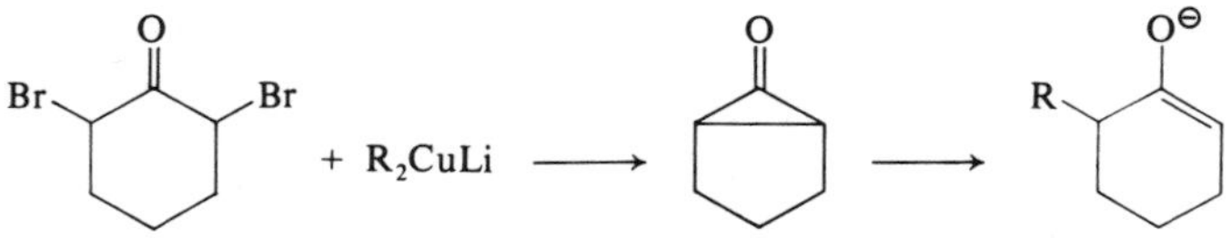

The semibenzilic pathway is solely operative during the Favorskii rearrangement of 2-bromocyclobutanone (7). The unusual hardness of the cyclobutanone carbonyl carbon certainly contributes to its electrophilicity toward hard donors such as water. The increased hardness of the small ring ketonic carbon is due to the lower p character of its bonds (exocyclic) to the oxygen atom. A similar ring contraction has been observed for 1,2-cyclobutanedione (8). Cyclobutanones undergo facile Baeyer–Villiger type oxidation by treatment with hydrogen peroxide (9) alone.

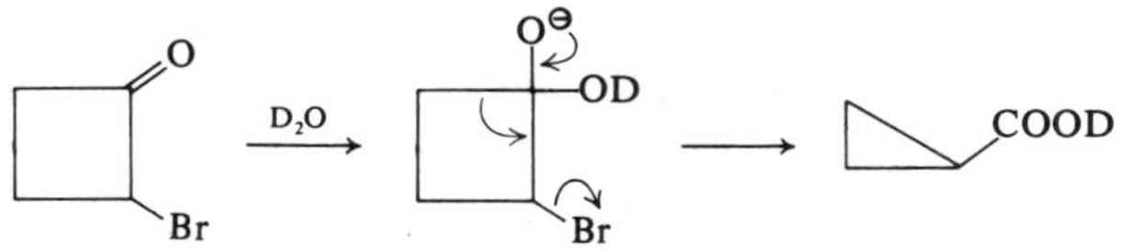

It should be mentioned that perfluoro- and perchlorocyclobutanones form stable α-halohydrins (10) in the same manner that chloral and α-ketoaldehydes do. Cyclopropanones (11) give hydrates, hemiacetals, and cyanohydrins extremely readily.

7.1.3. Miscellaneous Reactions

A large part of carbonyl chemistry is concerned with enolization. Kinetic deprotonation of ketones suggests the following preference: $CH_3CO > CH_2CO > CHCO$. On considering the carbanions as acid–base complexes it becomes clear that the primary complex is better stabilized than the secondary one, which is, in turn, more stable than the tertiary complex.

$$H^\ominus \rightarrow \ddot{C}(R)CO$$

$$H^\ominus \rightarrow \ddot{C}HCO \qquad\qquad R^\ominus \rightarrow \ddot{C}(R)CO$$

$$R^\ominus \rightarrow \ddot{C}HCO$$

Aldol condensation of cyclopentanone with isobutyraldehyde in methanol (12) gives erythro adducts. This is an α,α'-diastereogenic reaction. The Reformatsky reaction (13, 14) also yields predominantly erythro products.

The hard silicon atom forms strong bonds with oxygenated compounds. This property has been exploited in the direct preparation of cyanohydrin trimethylsilyl ethers (15–17). Strikingly the carbonyl group of p-benzoquinones can be protected by this method (18).

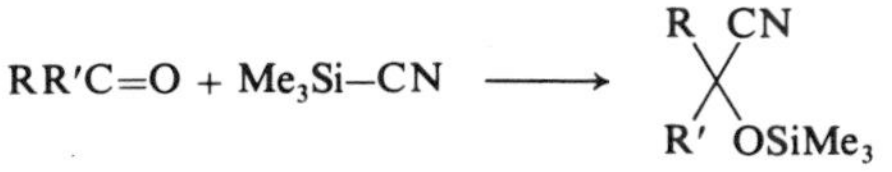

$$RR'C{=}O + Me_3Si{-}CN \longrightarrow \underset{R'\;\;OSiMe_3}{\overset{R\;\;CN}{\diagup\!\!\!\diagdown}}$$

A method for removal of halogen from α-bromoketones (19) utilizes a combination of LiI and BF_3 as the reagents. The hard–hard interaction between the carbonyl oxygen and the boron atom serves to activate the α-bromine toward attack by the soft iodide ion. The reaction is in accord with Saville's rule.

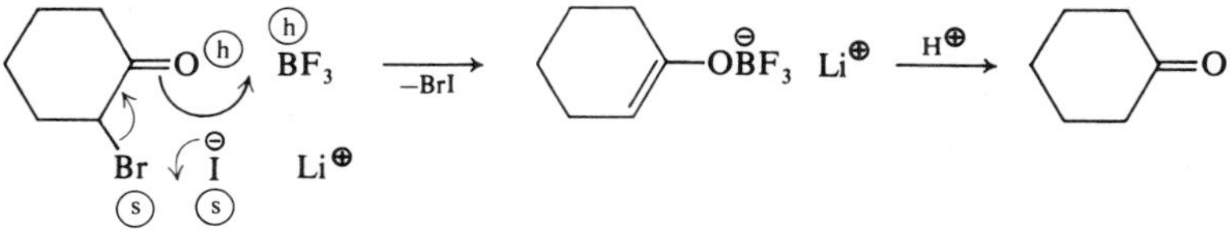

The unusual stereoisomer distribution (20) of the LiAlH$_4$ reduction products of 3-substituted bicyclo[2.2.2]octan-2-ones may be explained by assuming soft–soft nonbonded interactions between the C-3 substituent and the reducing agent in the transition state leading to the trans alcohol (Me, 50%; Et, 68%; i-Pr, 72%; Ph, 5%).

Concerning the stereochemistry of organometallic attack on cyclohexanones, it has been pointed out that axial approach is favored by harder nucleophiles (21).

7.2. THE CARBONYL OXYGEN AS A SOFT ACCEPTOR

The reductive coupling of carbonyl compounds with active metals (Na, Mg, Al) yields pinacols. An electron transfer from the metal surface to the carbonyl oxygen (ketyl formation), a soft–soft interaction, is undoubtedly involved. The conversion of esters to acyloins (22, 23) on the surface of metallic sodium is well known. Here the enediolate products can be trapped *in situ* by Me$_3$SiCl (24). The chlorosilane does not interfere with the coupling, yet it effectively removes the alkoxide ions and neutralizes the enediolate ions immediately on formation. The elimination of RO$^\ominus$ is imperative, for otherwise Claisen or Dieckmann condensations would compete with the normal course of reaction. These complicating processes require a hard base (e.g. RO$^\ominus$) to abstract a proton from the starting esters, whereas the desired coupling is accomplished by a soft base which is the electrons on the metal surface.

Imines are dimerized by sodium in an analogous fashion (25).

The reaction of diazomethane with α-diketones often leads to 1,3-dioxolines (26). It is plausible that attack on the carbonyl oxygen by the CH$_2$ of diazomethane is the initial step, for the zwitterion intermediates are somewhat stabilized.

In the light of this, it is interesting to note that the isatin derivatives afford mixtures of epoxides and 3-hydroxycarbostyrils, whereas the oxygen and sulfur heterocycles provide the dioxolines. The stabilization of the incipient carbanions generated during the addition of CH$_2$N$_2$ to the carbonyl oxygen by the neighboring lactone or thiolactone carbonyl is not shared by the isatin derivatives, as the lactam function is already polarized and can no longer serve as an electron sink. In other words, the keto carbonyl in isatin is not as polarizable as that in the O and S analogs.

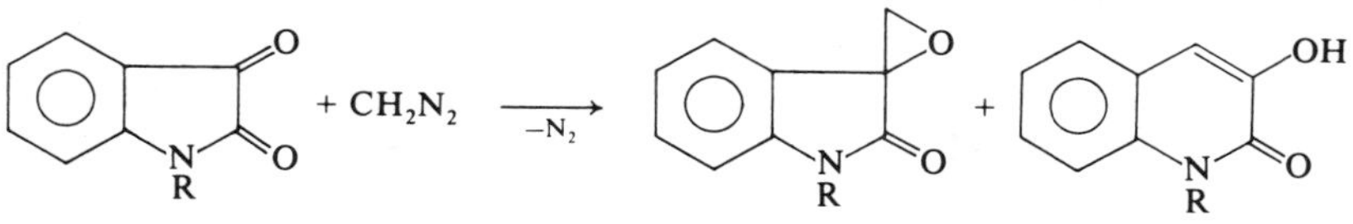

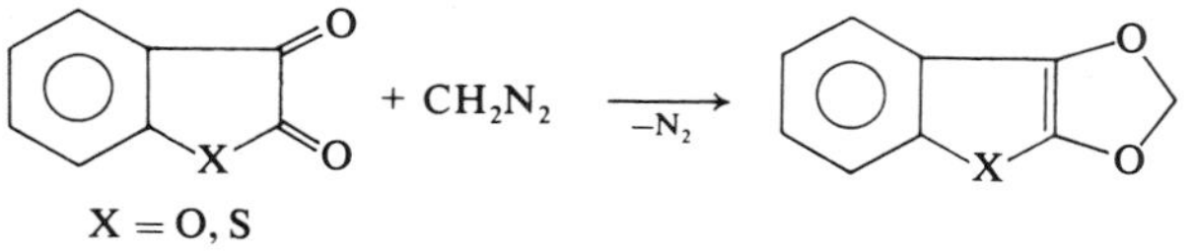

The formation of epoxy products from the reaction of diazomethane with α-keto esters and oxomalonates could also involve soft attack at carbonyl oxygen initially.

The Oppenauer oxidation of alcohols using fluorenone (27) as a hydride acceptor may not proceed by the mechanism in which $H^\ominus$ is delivered to the carbon of the carbonyl group. The following description remains a distinct possibility.

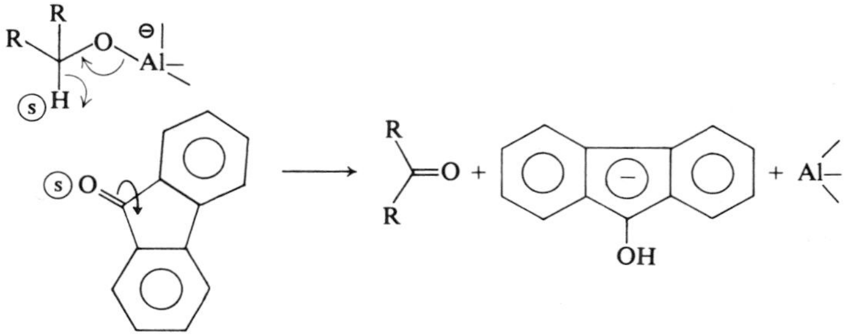

The dehydrogenation of certain alcohols by quinones (28) is thought to involve a hydride transfer.

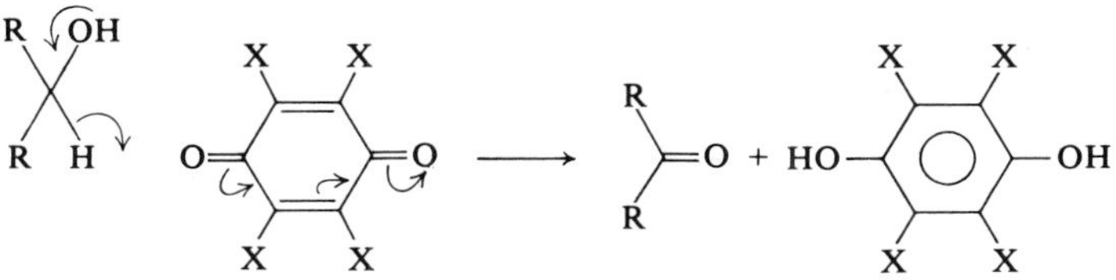

Cyanothioformates become accessible by the addition of thiols to carbonyl cyanide (29). Although the dicyanohydrin intermediates have been characterized, they may actually arise from a rearrangement of zwitterionic species containing an S—O bond.

$$\text{RSH} + \underset{\text{NC}}{\overset{\text{NC}}{\diagdown}}\text{C}{=}\text{O} \longrightarrow \text{RS}\overset{\text{H}}{\diagdown}\overset{\oplus}{\text{O}}\text{--}\overset{\ominus}{\text{C}}(\text{CN})_2$$

$$\downarrow$$

$$\underset{\underset{\text{CN}}{|}}{\overset{\overset{\text{OH}}{|}}{\text{RS--C--CN}}} \xrightarrow{\ \Delta\ } \text{RS}\overset{\text{O}}{\diagup}\overset{\|}{\diagdown}\text{CN} + \text{HCN}$$

Another method for the debromination of α-bromoketones (30) calls for the use of $NaBH_4$ in conjunction with heavy metal salts. The (S) $\cdots$ (S) $\sim\sim\sim$ (S) $\cdots$ (S) interactions in the transition state are the crucial requisite.

$$R\text{--}\underset{\underset{\text{R}'}{|}}{\overset{\overset{\text{O}}{\|}}{\text{C}}}\text{--CH--X} \quad M^{n\ \oplus} \longrightarrow R\overset{\text{OH}}{\diagup}\overset{}{\diagdown}\text{CHR}' \longrightarrow RCOCH_2R'$$

The thermal rearrangement of a ketosulfonium ylide (31) as depicted in the following equation proceeds via nucleophilic attack on the soft oxygen end of the carbonyl. A comparison of the mechanism with the Pummerer rearrangement is instructive.

$$\underset{\dot{\text{C}}\text{HPh}}{\overset{\text{O}}{\text{Ph}}}\diagdown\diagup\overset{\text{CH}_3}{\underset{\oplus}{\underset{\ominus}{\text{S}}}}\diagdown\text{CH}_3 \longrightarrow \underset{\dot{\text{C}}\text{HPh}}{\overset{\text{O}}{\text{Ph}}}\diagdown\diagup\overset{\ominus\text{CH}_2}{\underset{\oplus}{\text{S}}}\diagdown\text{CH}_3 \longrightarrow \underset{\dot{\text{C}}\text{HPh}}{\overset{\text{O--CH}_2\text{SMe}}{\text{Ph}}}\diagdown\diagup\diagdown\text{CH}_2$$

$$Me_2SO + Ac_2O \longrightarrow \underset{\text{Me}}{\overset{\text{Me}}{\diagdown}}\overset{\oplus}{\underset{}{\text{S}}}\text{--O}\text{--}\overset{\text{O}}{\underset{\|}{\text{C}}}\diagdown\text{Me}$$

$$OAc^{\ominus}$$

$$\downarrow \text{--HOAc}$$

$$\underset{\overset{|}{\overset{\ominus}{\text{C}}\text{H}_2}}{\overset{\text{Me}}{\diagdown}}\overset{\oplus}{\underset{}{\text{S}}}\text{--O}\diagup\overset{}{\underset{\text{O}}{\text{C}}}\diagdown\text{Me} \longrightarrow MeSCH_2OAc$$

7.3. REACTIONS OF α,β-UNSATURATED CARBONYL COMPOUNDS

7.3.1. Complex Metal Hydride Reductions

The reduction of α,β-unsaturated ketones with complex metal hydrides has been examined from the HSAB viewpoint (32). From an analysis based on the reasonable assumptions that in the conjugate enone system C-4 is softer than C-2 (carbonyl carbon) and that the more covalent M–H bond corresponds to the softer hydride, a unified picture emerges. The replacement of some of the hydride ions by hard alkoxy groups suppresses the softer conjugate addition to the enones. The proportion of 1,2- vs. 1,4-reduction of cyclopentenone varies from 14:86 using $LiAlH_4$ to a dramatic 90:9.5 using lithium trimethoxy-aluminum hydride (33). The α,β-unsaturated esters are reduced to allylic alcohols (34) by $LiAlH_4$ in the presence of ethanol. The ethoxyaluminum hydride is the active reducer.

Analogously, the reduction of cholestenone (35) with $NaBH_4$ and $NaBH(OMe)_3$ produces 1,2- and 1,4-reduction products in the ratios of 74:26 and 98:2, respectively.

The opposite effect is exerted by alkyl substituents. Potassium tri-sec-butyl-borohydride effects 1,4-reduction of cyclohexenones (36) exclusively.

Since boron is closer in electronegativity to hydrogen than aluminum, the B–H bond is more covalent than the Al–H bond. It follows that the borohydrides are softer than the corresponding aluminum hydrides. As a result, borohydrides are quite inert to protic solvents (hard $H^{\oplus}$ sources) and they tend to give more conjugate reduction products. The softer counterion $Na^{\oplus}$ also favors 1,4-reduction (cf. $NaBH_4$ vs. $LiBH_4$) (37). The alkali metal ions, in fact, play a significant role in modifying the substrate frontier orbital energy levels (38) through their interaction with the carbonyl oxygen. The extent of this modification is dependent on the hardness of the cation.

The addition of an amine to the metal hydrides limits the transfer of one hydride ion to the substrates. As the formation of alkoxyborohydride is also inhibited, 1,4-reduction is favored.

The conjugate reduction of enones is easier than that for enal (39, 40), because aldehyde carbonyl is softer than the ketone counterpart. Alkyl substituents in the α and β positions of the enones interfere with conjugate reduction (39). However, treatment of α-alkylthiocyclohexenones with $NaBH_4$ successfully gives saturated alcohols (41). It has been proposed that intramolecular $H^{\ominus}$ delivery from a S-coordinated borohydride is involved. A marked increase in the 1,4-reduction of enones by $LiAlH(SR)_3$ is observed (42) (see Table 7.1). Symbiotic softening of the reagents by the thio substituents is responsible for the reversal of the alkoxy effects.

TABLE 7.1   Percentage of 1,4–Reductions (42)

Enone	LiAlH$_4$	LiAlH$_4$ + 3MeOH	LiAlH$_4$ + 3t-BuOH	LiAlH$_4$ + 3MeSH	LiAlH$_4$ + 3t-BuSH
Cyclohexenone	22	5	78	56	95
Cyclopentenone	86	9.5	100	95	100

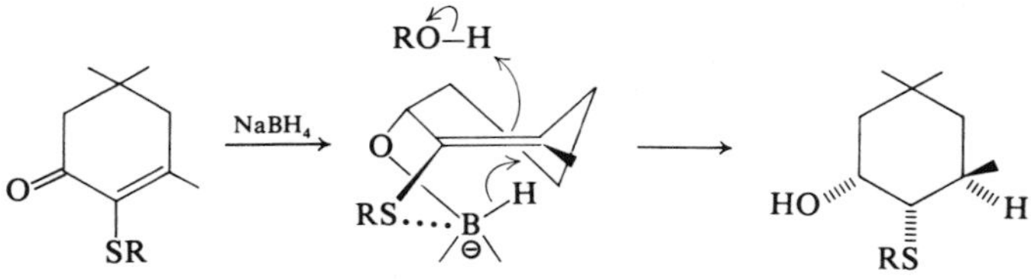

Cyclopentenone invariably gives more 1,4-addition products than cyclohexenone does in hydride reduction as well as in many other reactions. By *ab initio* calculations, it can be shown that the lowest unoccupied molecular orbital of cyclopentenone is 6.9 kcal lower in energy than cyclohexenone. As a consequence, the interaction at C-4 of cyclopentenone with soft bases is always stronger.

Aluminum hydrides in which the metal atoms do not bear a formal negative charge are harder; therefore aluminum hydride (34) and diisobutylaluminum hydride (43) attack selectively at the enone carbonyl.

1,4-Addition across the enone system is the major reaction course for organotin hydrides (44) which are quite soft.

7.3.2. Reactions with Organometallic Reagents

The hardness of the organometallic reagents (RLi > RMgX > RCu or R$_2$CuLi) is directly related to the metal atoms. In light of this, their reaction with enones is more readily understood.

Grignard reactions of 2-cyclooctenone (45) are shown below. The results are readily rationalized when one considers the hardness order: Me > Et > i-Pr > t-Bu.

Diorganocuprates effect Michael-type addition to α,β-enones. Although the reaction may or may not involve a one-electron transfer (46), the results are consistent with HSAB expectations.

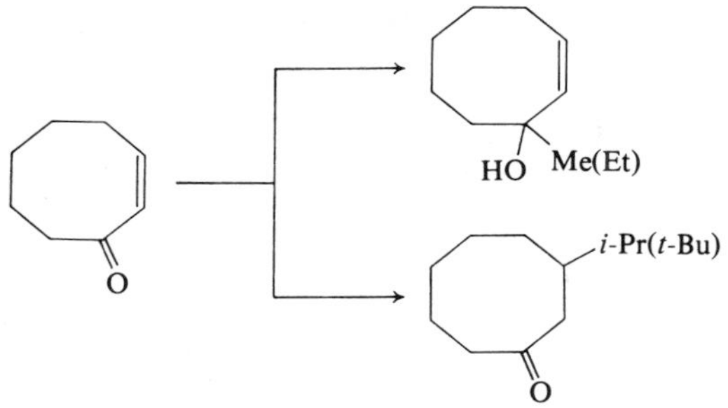

To avoid waste in synthesis, cuprates containing a nonmigrating group have been developed (47, 48). Among the stationary functions are the alkynyl, alkoxy, and phenylthio groups. The first two are harder than the other alkyl substituents, hence they are less prone to migrate to a soft center. The thio group does not interfere with the transfer of the other substituents because the S—Cu bond is very strong.

Not surprisingly, when only hard substituents are present in the cuprates, the reaction with α,β-unsaturated ketones takes a different course. Thus, lithium trialkynylcuprates add to the carbonyl selectively (49).

It has been claimed that α-chloroester carbanions and Grignard reagents derived from t-butyl esters of α-halocarboxylic acids are pyramidal and hard. They tend to attack the enone carbonyl. On the other hand, carbanions of α-chlorophenylacetates are planar, delocalized, and soft, and they behave as Michael donors (50).

For a long time the trityl anion has been used as a proton abstractor of weakly acidic substances. Its inertness toward the ketonic function has been regarded as steric in origin. It should be emphasized that the trityl anion is quite soft as demonstrated by the following reactions (51).

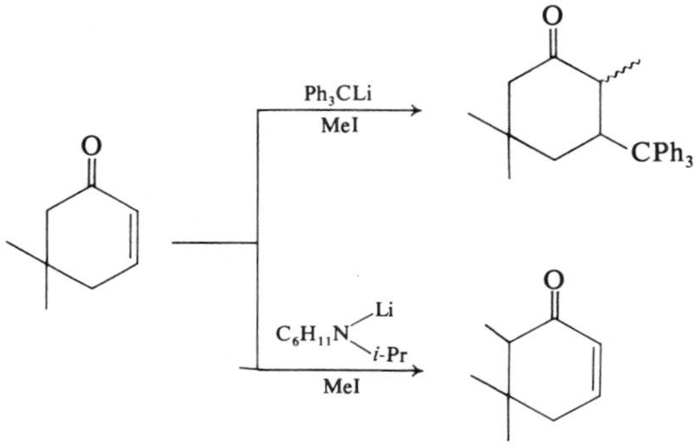

7.3.3. Reaction with Other Soft Bases

The 1,4-adducts are readily obtained when α,β-unsaturated carbonyl compounds are treated with methylthio(trimethyl)silane (52). The reaction conforms to Saville's rule as the product formation involves [s:s] $S:C_\beta$ and [h:h] $O:Si$ interactions.

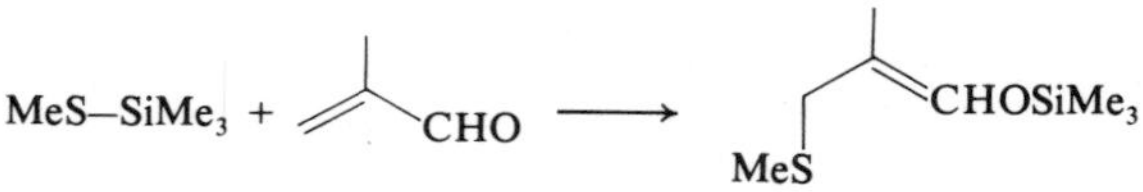

Nagata *et al.* have developed two procedures for hydrocyanation of α,β-unsaturated ketones (53, 54). The direct method using dialkylaluminum cyanides involves a fast equilibrium between the starting compounds and the 1,2-adducts and a slow reaction leading to 1,4-addition. The cyanide transfer is slow because $CN^\ominus$ in R_2AlCN is rather hard. On the other hand, the catalytic method (R_3Al–HCN) is an irreversible process. The oxygen of the enone is activated by a cationic species; the cyanide is delivered from the cyanoaluminate. As the transferring group becomes much softer in the anionic species $R_3AlCN^\ominus$, the 1,4-addition proceeds with great ease.

Dual pathways are available for the reaction of phosphonate anions with α,β-enones (55). The Michael addition is frontier orbital-controlled and is favored by the presence of proton sources, which quickly neutralize the charge on the enolate anions. These products decrease with time because they are gradually channeled into the charge-controlled Horner–Emmons olefination.

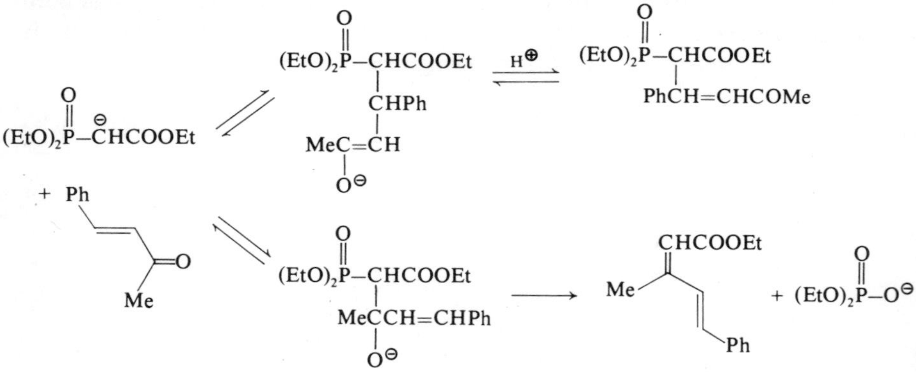

The α,β-enones possessing low energy LUMO behave according to this pattern.

7.3.4. Michael Additions

The 1,4-addition to α,β-unsaturated systems may be catalyzed by hard bases. Tertiary phosphines are also effective catalysts (56). Instead of directly removing a proton from the donor, they add to the unsaturated systems to generate the harder zwitterionic bases, which are more favorable for proton abstraction.

The attempt to add fluoride ion to α,β-unsaturated nitro compounds (57) was doomed to fail because of the extreme hardness of the halide. It has been demonstrated that $F^{\ominus}$ is an effective Michael catalyst (58–60) by virtue of its deprotonating ability.

Dinitromethide ions add to methyl acrylate in the Michael fashion. When the remaining hydrogen of the methide ions is replaced by fluorine, the Michael addition proceeds with considerable rate enhancement (61). The phenomenon has been attributed to anion destabilization through C–F bond weakening by the more electronegative trigonal carbon atom and the repulsion of the delocalized p electrons by the fluorine lone-pair electrons. The effect can simply be viewed as a consequence of the inharmonious union of the hard fluorine with the soft, delocalized carbanion center.

Acetylenedicarbonyl fluoride reacts with alcohols and amines to furnish esters and amides, respectively. On the other hand, ethanethiol adds to the triple bond of the substrate despite the extremely high reactivity of the acyl fluoride carbonyl (62). Aromatic amines undergo both Michael addition and amidation (63).

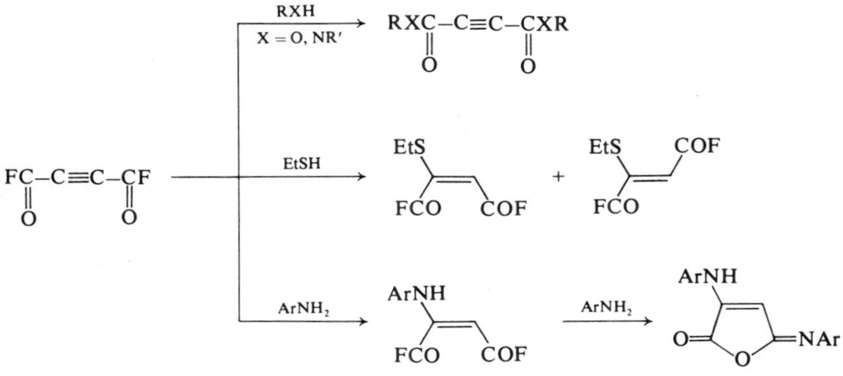

Many α,β-unsaturated ketones (p-quinones, in particular) add diazomethane to give diazoline adducts (26). The bonding processes at both ends involve soft–soft interactions. As Michael acceptors, p-benzoquinone and its sulfonylimine derivatives react with enamines to afford dihydrobenzofurans (64) and indolic compounds (65), respectively.

7.4. ASPECTS OF SOME CARBOXYLIC ACID DERIVATIVES

Peracids readily give up oxygen to soft bases. Thus, perbenzoic acid oxidizes diphenyldiazomethane to yield benzophenone (66), via an initial soft–soft interaction.

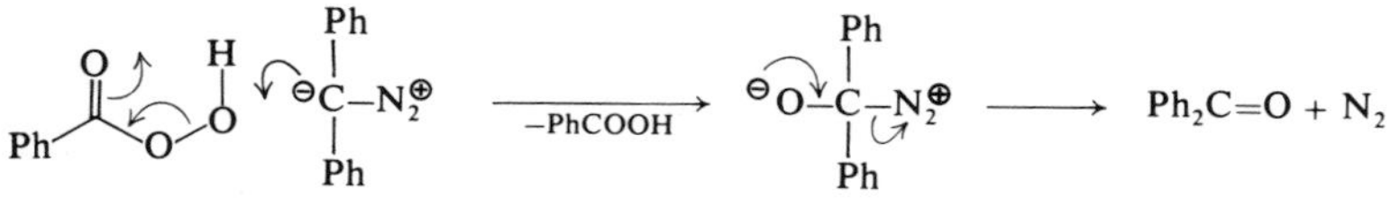

The reduction of valeryl peroxide with thiocyanate ion (67) presumably proceeds by a sequential soft interaction between S and O, and S and S.

$$(n\text{-BuCOO})_2 + KSCN$$
$$\downarrow$$
$$n\text{-BuCOOSCN} + n\text{-BuCOOK} \xrightarrow{\text{KSCN}} 2n\text{-BuCOOK} + (SCN)_2$$

Sodium perbenzoate is reduced by hydrogen peroxide. Radioactive tracer studies (68) suggest the following transition state with compatible reaction termini.

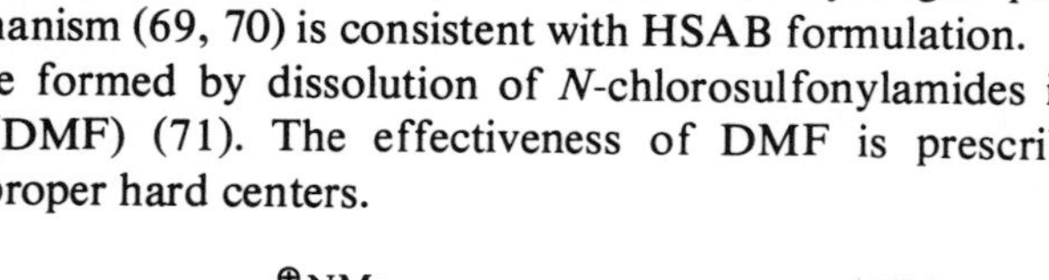

Nitriles are rapidly hydrated by treatment with hydrogen peroxide. The Wiberg mechanism (69, 70) is consistent with HSAB formulation.

Nitriles are formed by dissolution of N-chlorosulfonylamides in dimethylformamide (DMF) (71). The effectiveness of DMF is prescribed by the presence of proper hard centers.

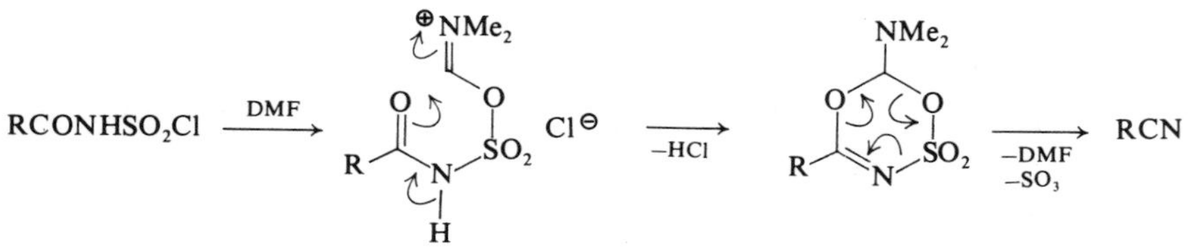

The conversion of amides into thioamides can be achieved by treatment with P_4S_{10}. This reaction exploits the intrinsically hard and oxaphilic nature of pentacovalent phosphorus which is bound to the sulfur atoms.

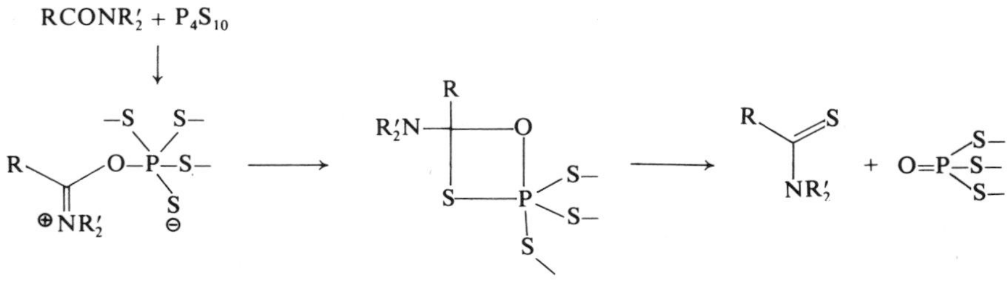

$$RCONR'_2 + P_4S_{10}$$

N,N-Di(2-picolyl)carboxamides are methanolyzed in the presence of cupric chloride (72). The complexation of the nitrogen atoms renders the amidic carbonyl much harder and better responsive to the hard base methanol.

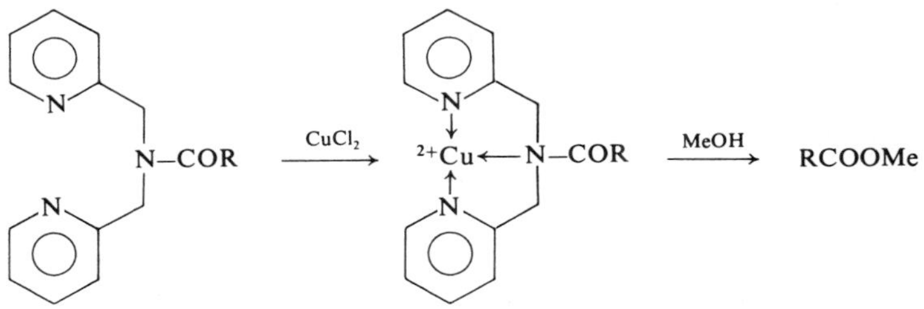

The choline selenol esters $RCOSeCH_2CH_2\overset{\oplus}{N}Me_3\ X^{\ominus}$ are susceptible to attack by thiols, yet they are stable to amines (73). In contrast, the choline thioesters are aminolyzed with ease. The gradated reactivities of the two types of esters are expected on the basis of symbiosis.

Saponification of certain esters such as *p*-nitrophenyl malonate and cyanoacetate proceeds via the E1cB mechanism (74). The α protons of these esters are very acidic (hard). *N*-Arylcarbamates also undergo alkaline hydrolysis (75) by the same route.

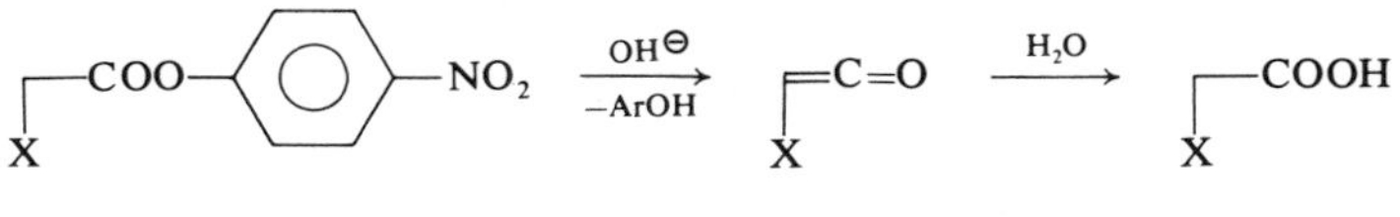

7.5. SOME CARBONYL-FORMING PROCESSES

In the absence of nucleophiles Lewis acids (e.g., BF_3) promote isomerization of epoxides to carbonyl compounds. The highly strained oxaspiropentanes are transformed into cyclobutanones (76) even by lithium salts or europium complexes. The essential feature is that a hard–hard interaction precedes bond migration.

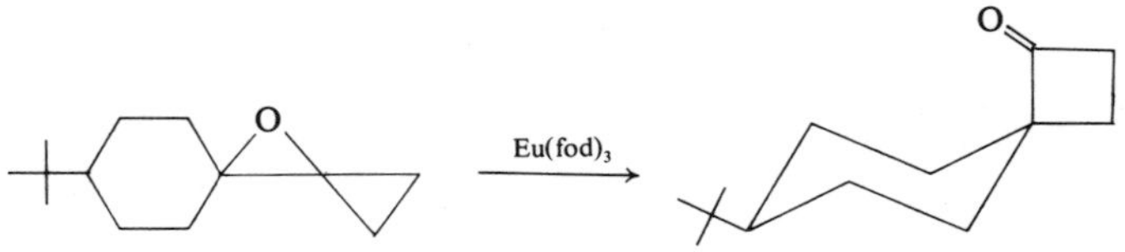

1,2-Diols can be dehydrogenated to α-ketols via their acetonides (77). The hydride acceptor is the relatively soft trityl cation.

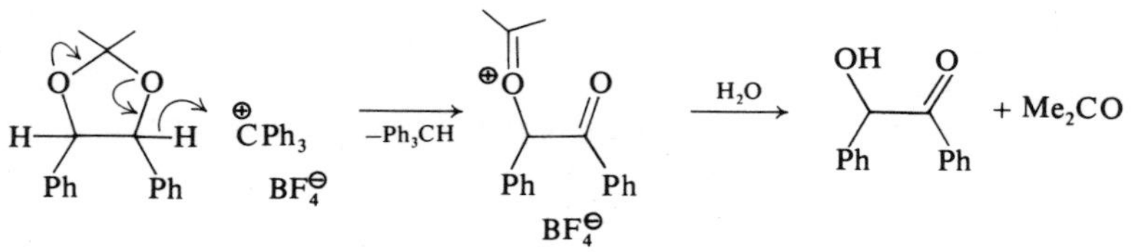

The oxidation of acetals to esters (78) may also involve hydride abstraction by the soft positive pole of the ozone molecule; the hard negative oxygen end occupies the site vacated by the hydride.

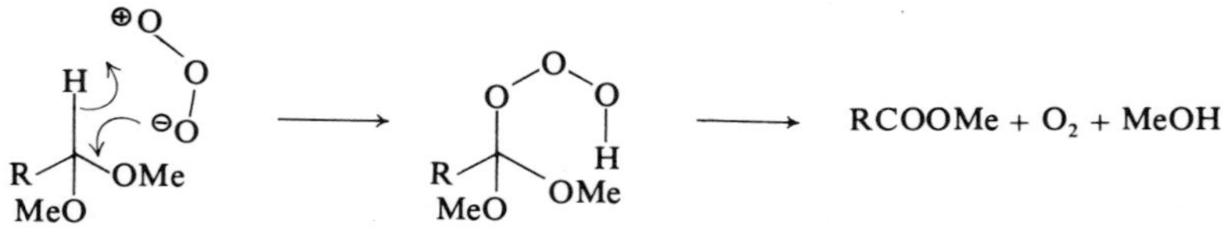

α-Hydroperoxycarboxylic acids undergo decarboxylative reduction to give the norketones (79) on exposure to dimethylformamide dimethylacetal. The reaction consists of the formation and fragmentation of a cyclic intermediate. In the most simplistic analysis of the facility and direction of such a fragmentation, only the soft–hard compatibility of the bond-forming partners need to be considered. For instance, the formation of a new $O-C(=O)$ bond between two hard atoms is essential for the eventual liberation of CO_2, and the carbonyl compound is generated through a soft–soft interaction in which the carbon acts as the donor and the oxygen as the acceptor.

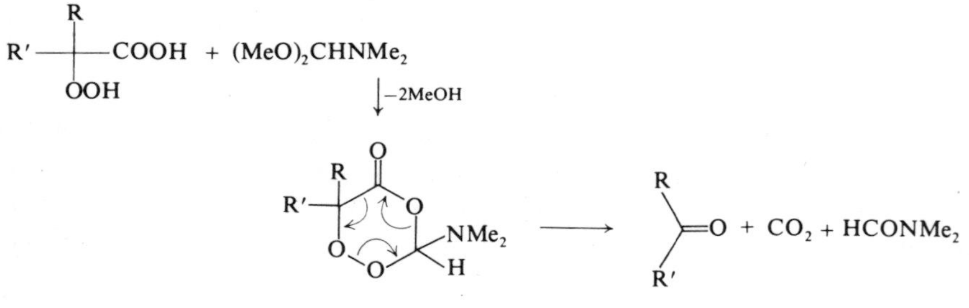

REFERENCES

1. F. Nerdel, P. Weyerstahl, and K. Lucas, *Tetrahedron Lett.* p. 5751 (1968).
2. C. H. Hassall, *Org. React.* **9**, 73 (1957).
3. R. R. Sauers, *Tetrahedron Lett.* p. 1015 (1962).
4. A. S. Kende, *Org. React.* **11**, 261 (1960).
5. R. B. Loftfield, *J. Am. Chem. Soc.* **73**, 4707 (1951).
6. G. H. Posner, C. E. Whitten, and J. J. Sterling, *J. Am. Chem. Soc.* **95**, 7788 (1973).
7. J.-M. Conia and J. Salaün, *Bull. Soc. Chim. Fr.* p. 1957 (1964).
8. J.-M. Conia and J. M. Denis, *Tetrahedron Lett.* p. 2845 (1971).
9. B. M. Trost, *Acc. Chem. Res.* **7**, 85 (1974).
10. S. Andreades and D. C. England, *J. Am. Chem. Soc.* **83**, 4670 (1961).
11. N. J. Turro, *Acc. Chem. Res.* **2**, 25 (1969).
12. J. E. Dubois and M. Dubois, *Chem. Commun.* p. 1567 (1968); cf. J. E. Dubois and P. Felimann, *Tetrahedron Lett.* p. 1225 (1975).
13. J. Canceill, J. J. Basselier, and J. Jacques, *Bull. Soc. Chim. Fr.* p. 1024 (1967).
14. F. Gaudemar-Bardone and M. Gaudemar, *C. R. Hebd. Seances Acad. Sci., Ser. C* **266**, 403 (1968).
15. D. A. Evans, L. K. Truesdale, and G. L. Carroll, *Chem. Commun.* p. 55 (1973).
16. D. A. Evans, G. L. Carroll, and L. K. Truesdale, *J. Org. Chem.* **39**, 914 (1974).
17. W. Lidy and W. Sundermeyer, *Chem. Ber.* **106**, 587 (1973).
18. D. A. Evans, J. M. Hoffman, and L. K. Truesdale, *J. Am. Chem. Soc.* **95**, 5822 (1973).
19. J. M. Townsend and T. A. Spencer, *Tetrahedron Lett.* p. 137 (1971).
20. D. Varech and J. Jacques, *Tetrahedron Lett.* p. 4443 (1973).
21. J. Huet, Y. Maroni-Barnaud, Nguyen Trong Anh, and J. Seyden-Penne, *Tetrahedron Lett.* p. 159 (1976).
22. S. M. McElvain, *Org. React.* **4**, 256 (1948).
23. K. T. Finley, *Chem. Rev.* **64**, 573 (1964).
24. U. Schräpler and K. Rühlmann, *Chem. Ber.* **96**, 2780 (1963).
25. J. Bastian and R. Jaunin, *Helv. Chim. Acta* **46**, 1248 (1963).
26. C. D. Gutsche, *Org. React.* **8**, 364 (1954).
27. E. W. Warnhoff and P. Reynolds-Warnhoff, *J. Org. Chem.* **28**, 1431 (1963).
28. L. M. Jackman, *Adv. Org. Chem.* **2**, 329–366 (1960).
29. M. T. Leplawy and A. Redliński, *Synthesis* p. 504 (1975).
30. T. Goto and Y. Kishi, *Tetrahedron Lett.* p. 513 (1961).
31. A. Terada and Y. Kishida, *Chem. Pharm. Bull.* **18**, 505 (1970).
32. J. Bottin, O. Eisenstein, C. Minot, and Nguyen Trong Anh, *Tetrahedron Lett.* p. 3015 (1972).
33. H. C. Brown and H. M. Hess, *J. Org. Chem.* **34**, 2206 (1969).
34. R. S. Davidson, W. H. H. Gunther, S. M. Waddington-Feather, and B. Lythgoe, *J. Chem. Soc.* p. 4907 (1964).
35. W. R. Jackson and A. Zurqiyah, *J. Chem. Soc.* p. 5280 (1965).
36. B. Ganem, *J. Org. Chem.* **40**, 146 (1975).
37. J. W. Wheeler and R. H. Chung, *J. Org. Chem.* **34**, 1149 (1969).
38. H. Handel and J. L. Pierre, *Tetrahedron* **31**, 2799 (1975).
39. M. R. Johnson and B. Rickborn, *J. Org. Chem.* **35**, 1041 (1970).
40. M. Mousseron, R. Jacquier, M. Mousseron-Canet, and R. Zagdoun, *Bull. Soc. Chim. Fr.* p. 1042 (1952).
41. J. W. deLeeuw, E. R. deWaard, P. F. Foeken, and H. O. Huisman, *Tetrahedron Lett.* p. 2191 (1973).
42. J. Durand, Nguyen Trong Anh, and J. Huet, *Tetrahedron Lett.* p. 2397 (1974).

43. K. E. Wilson, R. T. Seidner, and S. Masamune, *Chem. Commun.* p. 213 (1970).
44. M. Pereyre and J. Valade, *Bull. Soc. Chim. Fr.* p. 1928 (1967).
45. J. C. Damiano and A. Diara, *C. R. Hebd. Seances Acad. Sci. Ser. C* **276**, 441 (1973).
46. H. O. House and M. J. Umen, *J. Org. Chem.* **38**, 3893 (1973).
47. E. J. Corey and D. J. Beames, *J. Am. Chem. Soc.* **94**, 7210 (1972).
48. G. H. Posner, C. E. Whitten, and J. J. Sterling, *J. Am. Chem. Soc.* **95**, 7788 (1973).
49. G. Palmisano and R. Pellegata, *Chem. Commun.* p. 892 (1975).
50. Y. Maroni-Barnaud, M. C. Roux-Schmitt, and J. Seyden-Penne, *Tetrahedron Lett.* p. 3129 (1974).
51. R. A. Lee and W. Reusch, *Tetrahedron Lett.* p. 969 (1973).
52. D. A. Evans, K. G. Grimm, and L. K. Truesdale, *J. Am. Chem. Soc.* **97**, 3229 (1975).
53. W. Nagata, M. Yoshioka, and S. Hirai, *J. Am. Chem. Soc.* **94**, 4635 (1972).
54. W. Nagata, M. Yoshioka, and M. Murakami, *J. Am. Chem. Soc.* **94**, 4644 (1972).
55. B. Deschamps, N. T. Anh, and J. Seyden-Penne, *Tetrahedron Lett.* p. 527 (1973).
56. D. A. White and M. M. Baizer, *Tetrahedron Lett.* p. 3597 (1973).
57. S. Hoz, M. Albeck, and Z. Rappoport, *Synthesis* p. 162 (1975).
58. E. LeGoff, *J. Am. Chem. Soc.* **84**, 3975 (1962).
59. A. Ostaszynski, J. Wielgat, and T. Urbanski, *Tetrahedron* **25**, 1929 (1969).
60. I. N. Rozhkov and I. L. Knunyants, *Dokl. Akad. Nauk SSSR* **199**, 614 (1971).
61. L. A. Kaplan and H. B. Pickard, *Chem. Commun.* p. 1500 (1969).
62. F. E. Herkes and H. E. Simmons, *J. Org. Chem.* **40**, 420 (1975).
63. F. E. Herkes, *J. Org. Chem.* **40**, 423 (1975).
64. K. C. Brannock, R. D. Burpitt, H. E. Davis, H. S. Pridgen, and J. G. Thweatt, *J. Org. Chem.* **29**, 2579 (1964).
65. M. E. Kuehne, *J. Am. Chem. Soc.* **84**, 837 (1962).
66. R. Curci, F. DiFuria, and F. Marcuzzi, *J. Org. Chem.* **36**, 3774 (1971).
67. C. L. Jenkins and J. K. Kochi, *J. Org. Chem.* **36**, 3053 (1971).
68. K. Akiba and O. Simamura, *Tetrahedron* **26**, 2519 and 2527 (1970).
69. K. B. Wiberg, *J. Am. Chem. Soc.* **77**, 2519 (1955).
70. J. E. McIsaac, Jr., R. E. Ball, and E. J. Behrman, *J. Org. Chem.* **36**, 3048 (1971).
71. G. Lohaus, *Chem. Ber.* **100**, 2719 (1967).
72. R. P. Houghton and R. R. Puttner, *Chem. Commun.* p. 1270 (1970).
73. A. Makriyannis, W. H. H. Gunther, and H. G. Mautner, *J. Am. Chem. Soc.* **95**, 8403 (1973).
74. T. C. Bruice and B. Holmquist, *J. Am. Chem. Soc.* **90**, 7136 (1968).
75. A. Williams, *J. Chem. Soc., Perkin Trans. 2*, 808 (1972).
76. B. M. Trost and M. J. Bogdanowicz, *J. Am. Chem. Soc.* **95**, 2038 (1973).
77. D. H. R. Barton, P. D. Magnus, G. Smith, and D. Zurr, *Chem. Commun.* p. 861 (1971).
78. C. Moreau and P. Deslongchamps, *Can. J. Chem.* **49**, 2465 (1971).
79. H. H. Wasserman and B. H. Lipshutz, *Tetrahedron Lett.* p. 4611 (1975).

8

Organophosphorus Reactions

8.1. GENERAL CONSIDERATIONS

The majority of phosphorus compounds can be divided into three classes: trivalent, positively charged, and pentacovalent compounds.

The predominant role of trivalent phosphorus is that of a nucleophile. Phosphines are the simplest derivatives and their basicity decreases according to substitution: $R_3P > R_2PH > RPH_2 \gg PH_3$. As $H^\ominus$ is much softer than the alkyl residue, the replacement of H by R would make the phosphorus atom harder and, hence, easier to combine with the proton.

For R_3P, the nucleophilicity diminishes (1) along the series R = alkyl > aryl > alkoxy, and within the phosphite subgroup, R = MeO > EtO > n-PrO > n-BuO. The progressive increase in hardness of the alkyl, aryl, and alkoxy groups makes the phosphorus correspondingly harder and less compatible with a soft acid. The fact that the trialkylphosphites are still reactive toward alkyl halides (Michaelis–Arbuzov reaction) may be rationalized in the following manner. The ground state energies of phosphites must be higher than phosphines by virtue of the negative contribution by the hard–soft interactions between the O and P atoms. However, on going into the transition state of alkylation the phosphorus atom undergoes rehybridization and a valence change which can essentially be summarized as a soft to hard transformation. The activation energies for such a process must be lower. Thus, both the elevated ground state energy and the decreased activation energy contribute to the ease of phosphite alkylation.

Before discussing the various aspects of phosphorus chemistry, one should keep in mind that the trivalent phosphorus compounds frequently exhibit biphilicity (2), as shown in the following reactions.

$$R_3P + :CR_2' \longrightarrow R_3\overset{\oplus}{P}-\overset{\ominus}{C}R_2' \longleftrightarrow R_3P=CR_2'$$

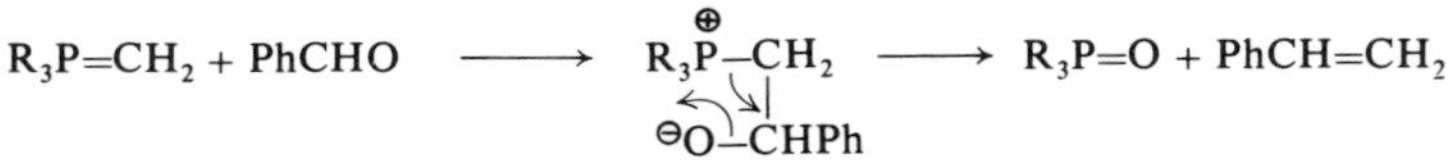

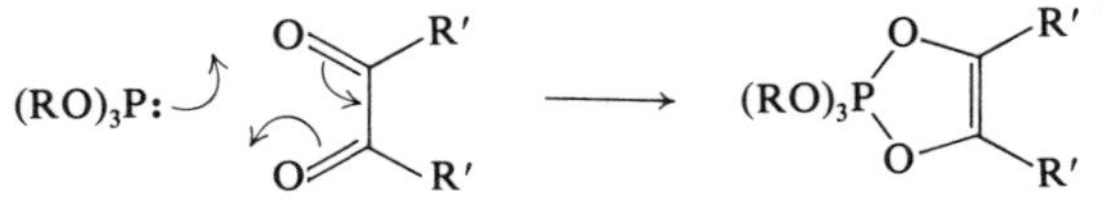

$$R_3P=CH_2 + PhCHO \longrightarrow R_3\overset{\oplus}{P}-CH_2 \longrightarrow R_3P=O + PhCH=CH_2$$
$$\overset{\ominus}{O}-CHPh$$

The chemistry of pentacovalent and charged tetrahedral phosphorus compounds is equally fascinating and biochemically significant. In various reactions the central phosphorus always acts as a hard Lewis acid.

8.2. TRIVALENT PHOSPHORUS COMPOUNDS AS NUCLEOPHILES

8.2.1. Attack at Carbon

Tertiary phosphines are rapidly quaternized by alkyl halides. During the reaction with the ambient triethoxycarbenium fluoroborate, phosphines seek the softer ethyl group (3).

$$R_3P + (EtO)_3C^{\oplus} BF_4^{\ominus} \longrightarrow R_3\overset{\oplus}{P}Et\ BF_4^{\ominus} + (EtO)_2CO$$

Phosphines function as Michael donors in the betaine formation (4, 5) with benzoquinone. Similarly, ylides are produced from mixing phosphines with maleic anhydride (6, 7) and benzyne (8).

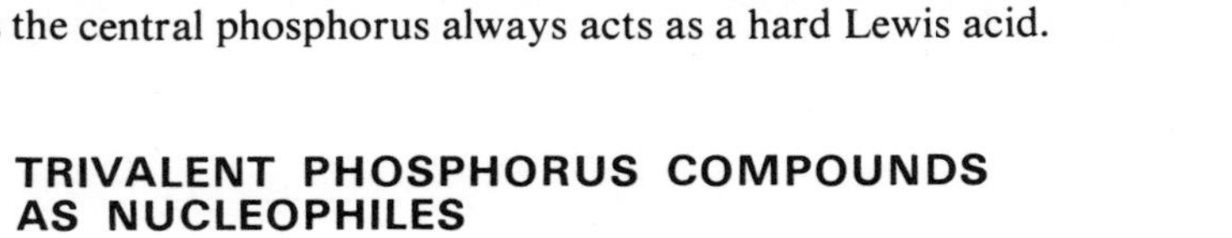

$$Ph_2PMe + \quad \longrightarrow \quad \longrightarrow Ph_3P=CH_2$$

The ring expansion of 9-methyl-9-phosphafluorene on treatment with methyl propiolate (9) must be initiated by a Michael addition of the phosphine to the triple bond.

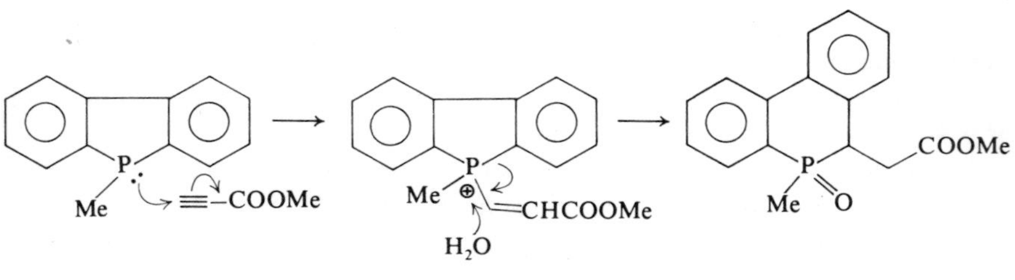

By virtue of their softness, phosphines readily intercept carbenes resulting in Wittig reagents (10).

$$Ph_3P + CH_2Cl_2 + n\text{-BuLi} \longrightarrow Ph_3P=CHCl + n\text{-BuH} + LiCl$$

In the Perkov reaction (11), all evidence indicates that the addition of the phosphite to the carbonyl carbon is the first step. The Perkov pathway is favored over the competing Michaelis–Arbuzov reaction (attack on X) by the harder halide substrates (12), e.g., $RCOCH_2Cl > RCOCH_2Br > RCOCH_2I$.

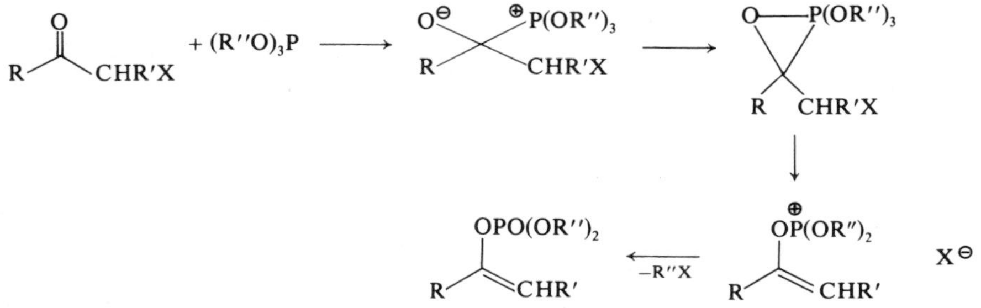

α-Phenylthiodeoxybenzoin reacts with $(EtO)_3P$ following the Perkov route, in contrast to the α-ethylthio analog which furnishes a host of ketonic and vinylic products (13). It appears that, besides steric effects, the slight increase in hardness of the phenylthio sulfur drives the phosphorus atom to interact with the carbonyl.

In the Perkov reaction of α-chlorothioacetone (14), decomposition of the intermediate zwitterion is determined by the softness of the sulfide anion.

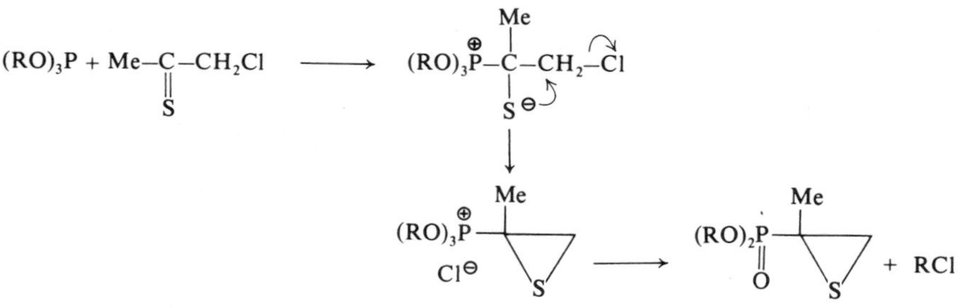

The Arbuzov reaction of substituted α-bromoacetophenones shows a negative Hammett ρ value suggesting an initial attack on carbon during formation of the ketophosphonates (15).

Although phosphines seldom attack simple carbonyl compounds, Mark (16) reported the formation of zwitterionic adducts from hexamethylphosphorous triamide with aldehydes. Because the aldehydic carbon is softer than the ketone counterpart and the phosphorus of the triamide is harder than that of ordinary phosphines, the softness of these reactants are perhaps sufficiently compatible with each other.

Sodium diethylphosphonate combines with ω-chloroalkanols of appropriate chain length to yield cyclic phosphonates (17). It is evident that the first P-alkylation step is a soft–soft combination and that the ring closure involves a hard–hard interaction.

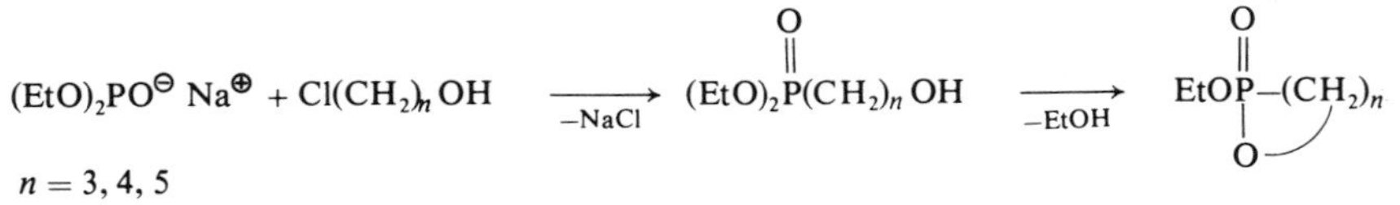

$$(EtO)_2PO^{\ominus}\,Na^{\oplus} + Cl(CH_2)_n\,OH \xrightarrow[-NaCl]{} (EtO)_2\overset{\overset{\displaystyle O}{\|}}{P}(CH_2)_n\,OH \xrightarrow[-EtOH]{} EtO\overset{\overset{\displaystyle O}{\|}}{\underset{\underset{\displaystyle O}{|}}{P}}{-}(CH_2)_n$$

$$n = 3, 4, 5$$

The displacement reaction of mesylates by $(EtO)_2PONa$ has been investigated (18). Apart from the normal Michaelis–Becker products (P-alkylation), a substantial amount of olefins and O-alkylation compounds are detected in polar solvents. The mesyloxy group is more readily displaced by the phosphonate oxyanion as it is harder than halides. The olefins arise from a hard-type process.

A related report (19) reveals that phosphites are generated by the reaction of triarylchloromethanes with silver diisopropylphosphite. This reaction is charge-controlled.

The vinyl phosphonates act as Michael acceptors of phosphonate ions (20). The soft P atom of the latter species is utilized for the bond formation.

The kinetic product of sodium diphenylphosphite addition to tetraphenyl-cyclopentadienone is a 1,6-adduct (21).

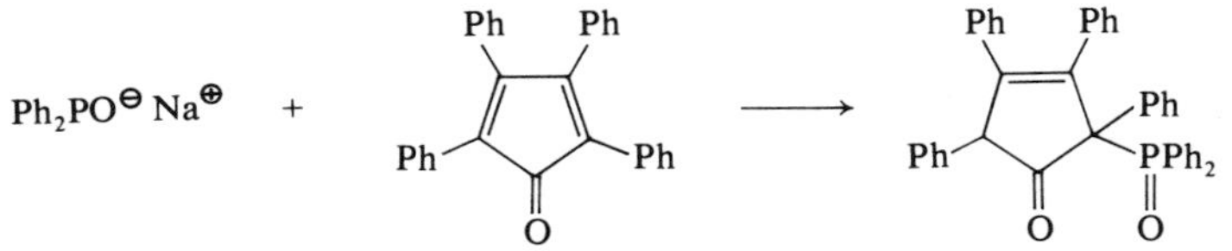

8.2.2. Attack at Halogen

The direct attack on the halogen of α-halocarbonyl compounds by a tertiary phosphine is implicated by substrate reactivity indices (22). The change of venue from the Perkov and Arbuzov pathways concurs with the HSAB rationale as the phosphine P is softer than the phosphite P. Studies on enol phosphonium salt formation (23) support this contention.

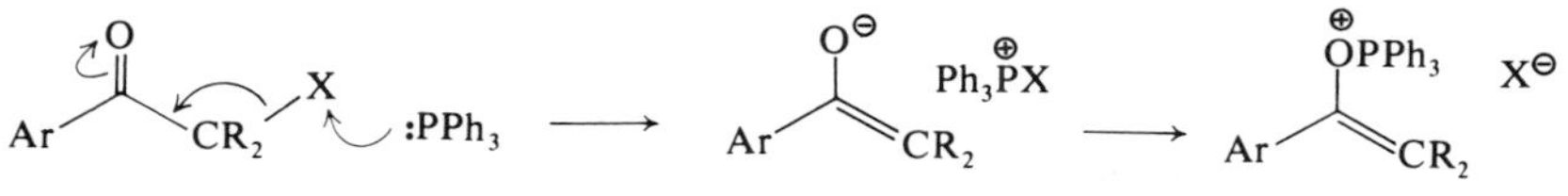

Dehalogenation of α-bromoacetophenones and α-bromopropiophenones with or without general acid catalysis can be effected by triphenylphosphine (24) but not by trialkylphosphites. The α-chloroketones fail to undergo a similar reaction. The catalyzed reaction conforms to Saville's rule.

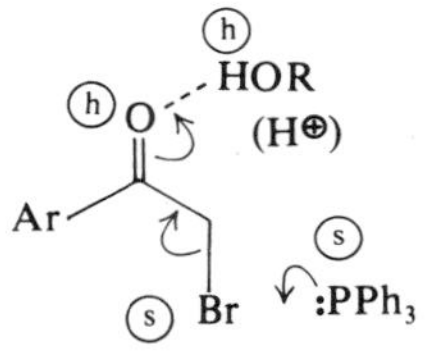

The C-4 bromine of 2,4,6-tribromo-4-methylcyclohexa-2,5-dienone is very soft, and therefore it is readily extracted by trivalent phosphorus compounds (25). Rate studies with the 2,6-di-*tert*-butyl analog exclude the attack on oxygen (26).

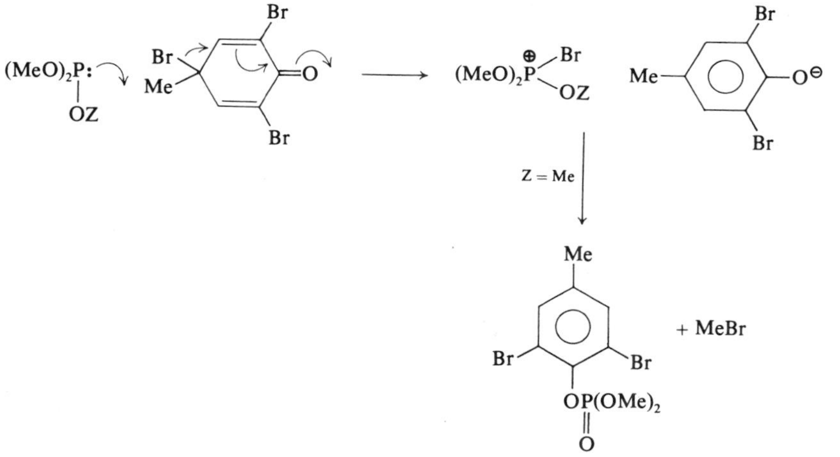

The cyclohexadienone (imine) tautomers of *o*- and *p*-halophenols (and anilines) are most likely the reactive species during their dehalogenation (27) at elevated temperatures.

Dibromoketene can be generated from trimethylsilyl tribromoacetate (28) on treatment with Ph₃P. Many other trihaloacetic acid derivatives undergo similar dehalogenation reactions.

$$Ph_3P: \quad Br-CBr_2-\overset{\overset{O}{\parallel}}{C}-OSiMe_3 \quad \longrightarrow \quad Br_2C=C=O$$

It is interesting to note that N,N-diethylfluorodichloroacetamide is less reactive than trichloroacetamide toward triphenylphosphine (29), because of the destabilization of an α-carbanion by the hard fluorine.

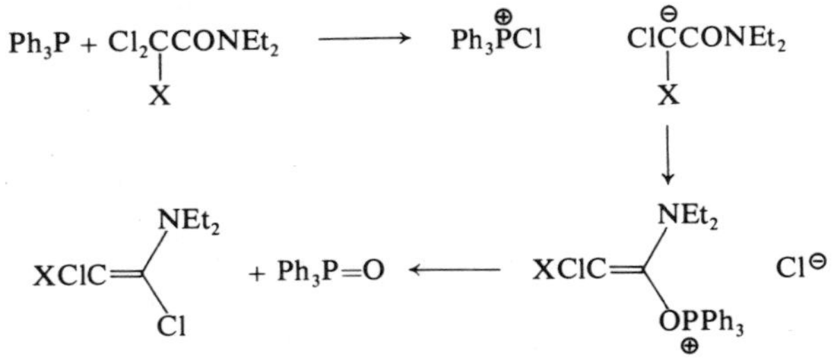

$$Ph_3P + \underset{\underset{X}{|}}{Cl_2}CCONEt_2 \quad \longrightarrow \quad Ph_3\overset{\oplus}{P}Cl \quad \underset{\underset{X}{|}}{Cl\overset{\ominus}{C}}CCONEt_2$$

$$XClC\!\!=\!\!\overset{\diagup NEt_2}{\diagdown Cl} \quad + Ph_3P\!=\!O \quad \longleftarrow \quad XClC\!\!=\!\!\overset{\diagup NEt_2}{\underset{\underset{\oplus}{OPPh_3}}{\diagdown}} \quad Cl^{\ominus}$$

It is plausible that the following reactions are biphilic rather than direct substitutive reactions. Both hardness and steric factors favor N–P bond formation in the last example.

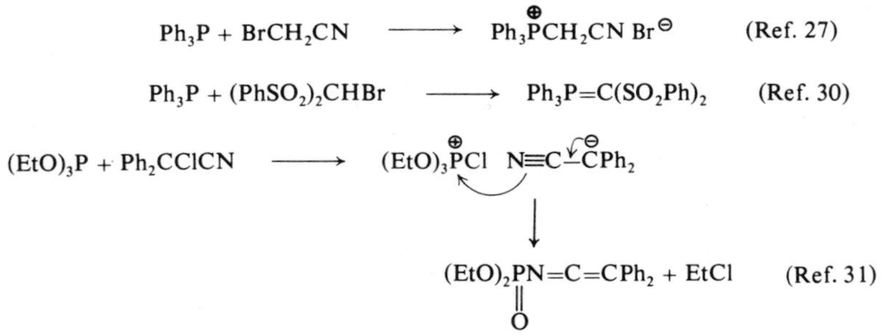

$$Ph_3P + BrCH_2CN \quad \longrightarrow \quad Ph_3\overset{\oplus}{P}CH_2CN \ Br^{\ominus} \qquad (Ref. 27)$$

$$Ph_3P + (PhSO_2)_2CHBr \quad \longrightarrow \quad Ph_3P\!=\!C(SO_2Ph)_2 \qquad (Ref. 30)$$

$$(EtO)_3P + Ph_2CClCN \quad \longrightarrow \quad (EtO)_3\overset{\oplus}{P}Cl \ N\!\!\equiv\!\!C\overset{\ominus}{\smile}CPh_2$$

$$(EtO)_2PN\!=\!C\!=\!CPh_2 + EtCl \qquad (Ref. 31)$$
$$\underset{O}{\overset{\parallel}{}}$$

The haloacetylenes afford phosphonium salts or phosphonates when reacted with phosphines or phosphites, respectively (32–36). However, in the presence of a proton source, a net reduction to the acetylenes is observed.

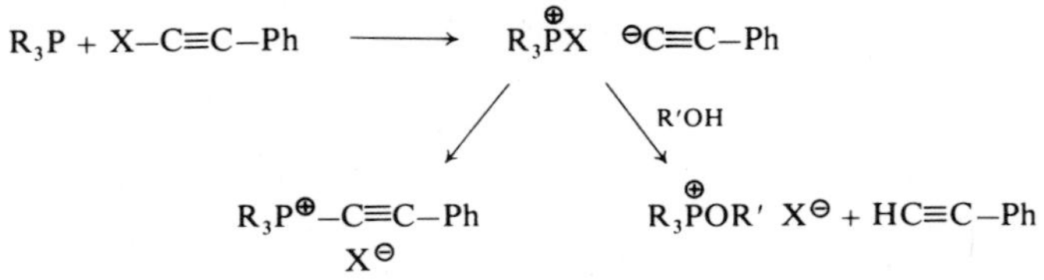

$$R_3P + X-C\!\!\equiv\!\!C-Ph \quad \longrightarrow \quad R_3\overset{\oplus}{P}X \ ^{\ominus}C\!\!\equiv\!\!C-Ph$$

$$R_3P^{\oplus}\!\!-\!C\!\!\equiv\!\!C-Ph \qquad\qquad R_3\overset{\oplus}{P}OR' \ X^{\ominus} + HC\!\!\equiv\!\!C-Ph$$
$$X^{\ominus} \qquad\qquad\qquad\qquad\qquad R'OH$$

Sodium diphenylphosphide reacts with 3-bromopropyne in liquid ammonia to furnish propyne and aminodiphenylphosphine (37), even though 3-

bromopropyne is known to undergo facile ammonolysis. The interaction between the softest species is overwhelmingly preferred. However, a tertiary
phosphine is formed in a similar reaction with 3-chloropropyne. The
propargylic carbon is apparently softer than the chlorine.

$$Ph_2P^{\ominus} \, Na^{\oplus} + BrCH_2C\equiv CH \longrightarrow Ph_2PBr + Na^{\oplus} \; {}^{\ominus}CH_2C\equiv CH$$

$$\downarrow NH_3$$

$$Ph_2PNH_2 + MeC\equiv CH + NaBr$$

The debromination of activated *vic*-dibromides by trivalent phosphorus compounds has been demonstrated. The process is an E2 type (38, 39). The more
potent dicyclohexylphosphide anion is capable of debrominating 1,2-
dibromoethane (40). It is of interest to note that the slightly harder diphenylphosphide elects to attack on the carbon atom (S_N2).

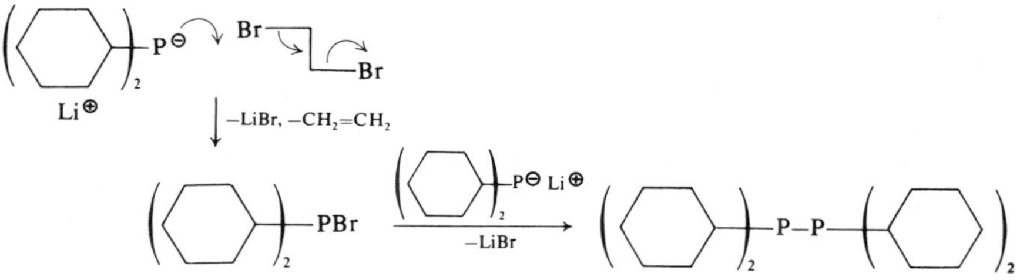

Perchlorofulvalene (41) and perchlorofulvene (42) have been synthesized by
the dehalogenation of decachlorobi-2,4-cyclopentadien-1-yl and 5-trichloro-
methylperchlorocyclopentadiene, respectively, with triethylphosphite.
Tetrahalomethanes react with phosphines in a 1 : 2 molar ratio to afford
dihalomethylenephosphoranes and dihalophosphoranes (43, 44). 1,2-Bis-
(phenylphospha)cyclohexane is cleaved by carbon tetrachloride (45).

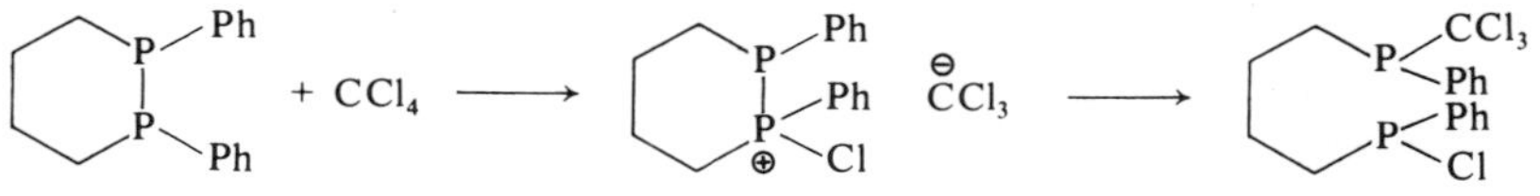

A mixture of a dialkyl hydrogen phosphonate, triethylamine, and carbon
tetrachloride is very effective for dehydration of aldoximes (46). The actual
reagent is formed by the deprotonation of the phosphonate (phosphite

tautomer) by the hard amine, followed by $Cl^{\ominus}$ removal from CCl_4 by the ensuing phosphonate ion.

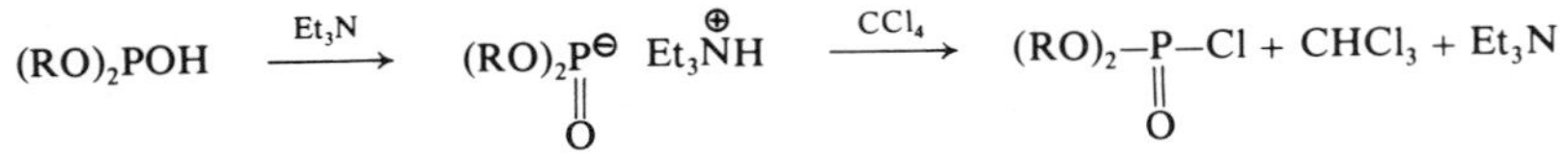

$$(RO)_2POH \xrightarrow{Et_3N} (RO)_2\overset{\ominus}{\underset{\underset{O}{\parallel}}{P}} Et_3\overset{\oplus}{N}H \xrightarrow{CCl_4} (RO)_2\underset{\underset{O}{\parallel}}{-P-Cl} + CHCl_3 + Et_3N$$

Whereas the strong hard bases deprotonate alkyltriphenylphosphonium salts to generate the ylides, triphenylphosphine tends to abstract the soft bromine from bromomethyltriphenylphosphonium halides and the bromodifluoromethyl analog (47, 48). It is pertinent to mention the reaction of phenyllithium with halomethylphosphonium salts (49). The chlorometho salt undergoes deprotonation only, while the bromometho and iodometho analogs give a mixture of ylides in which the halogen atom is either retained or lost. The higher proportion of $Ph_3P=CH_2$ obtained from the iodomethyl compound, compared to the bromomethyl derivative, manifests the dual properties of organolithium compounds as being both basic and soft.

$$Ph_3P + Ph_3P^{\oplus}-CZ_2Br\ X^{\ominus} \longrightarrow Ph_3P=CZ_2 + Ph_3\overset{\oplus}{P}Br\ X^{\ominus}$$

$$Z = H, F$$

As anticipated, the N-halosuccinimides (50) and alkyl hypohalites are reactive toward trivalent phosphorus compounds. The biphilic mechanism for hypohalite decomposition has been clarified using optically active phosphines (51, 52).

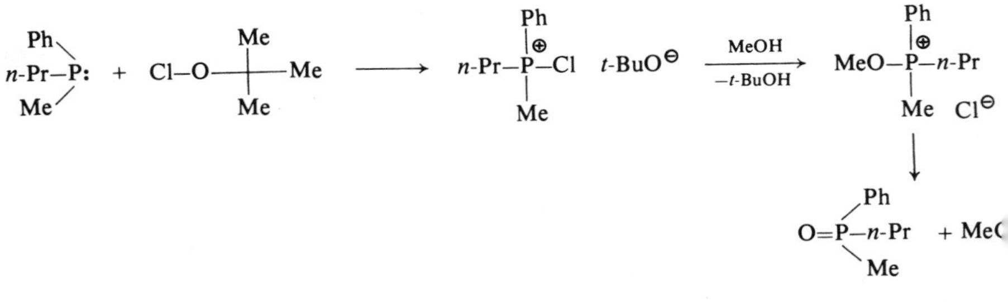

8.2.3. Attack at Oxygen

The peroxy compounds have high energy contents. The weakness of the O—O bond (35 kcal/mole) (53) is a reflection of the unfavorable $O^{\ominus}$ (hard) and $O^{\oplus}$ (soft) combination. Diacyl peroxides, for example, react vigorously with soft donors such as phosphines (54, 55). This deoxygenation process has been applied to the synthesis of the highly strained malonic anhydrides (56).

The fragmentation of β-peroxylactones promoted by phosphines yields two types of products (57). Diisopropyl peroxydicarbonate is similarly degraded (58). Here the phosphorane intermediate decomposes into a carbonate and a pyrocarbonate via different pathways.

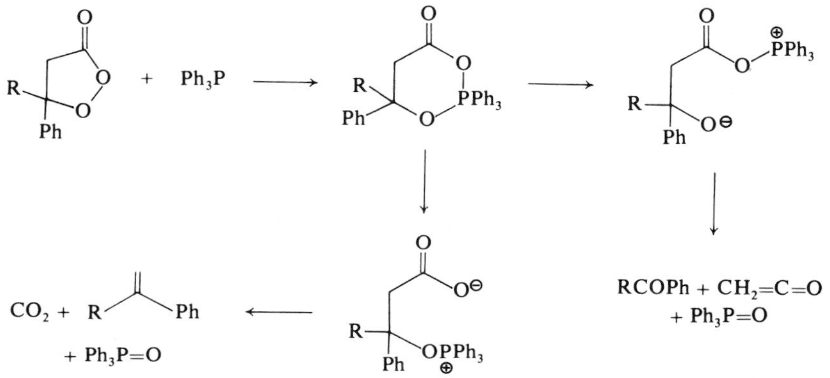

In contradistinction to the phosphine–peroxy compound reaction, an ionic mechanism is not operative in the corresponding amine reaction. The hard amine bases are less prone to interact with the soft oxy center, rendering the alternative one-electron transfer process viable (59, 60).

The existence of pentacovalent phosphorus intermediates during the deoxygenation of dialkyl peroxides with phosphites has been demonstrated by ^{31}P nuclear magnetic resonance studies (61) and isotope labeling techniques (62, 63). The corresponding adducts derived from phosphines are less stable because they lack full symbiosis.

Depending on the structure of the p-quinones and the nature of the trivalent phosphorus compounds being used, two types of products may result from their reactions. The initial attack is directed at either the O or the C of the quinones (64, 65), and both sites are soft.

Pentacovalent phosphorus cyclic adducts are produced by the addition of o-quinones and α-diketones to phosphites. The structure of one such biphilic adduct from phenanthraquinone has been determined by X-ray diffraction (66).

As previously mentioned, cyclopentadienones which possess a very soft oxygen acceptor site react vigorously with soft donors. Although stable adducts of these compounds with phosphines (67, 68) have been isolated, their reaction with phosphites (69, 70) is more complex. However, a soft–soft interaction is involved in each step.

Phthalic anhydride is transformed into diphthalyl (71) on heating with triethylphosphite. Diphenylketene undergoes deoxygenative rearrangement by the same treatment (72). The analogous reaction of isocyanates furnishes isonitriles. In this case the intermediate disintegrates into a "carbene" which is stabilized by a neighboring nitrogen atom.

$$Ph_2C=C=O + (EtO)_3P \longrightarrow PhC\equiv CPh + (EtO)_3P=O$$

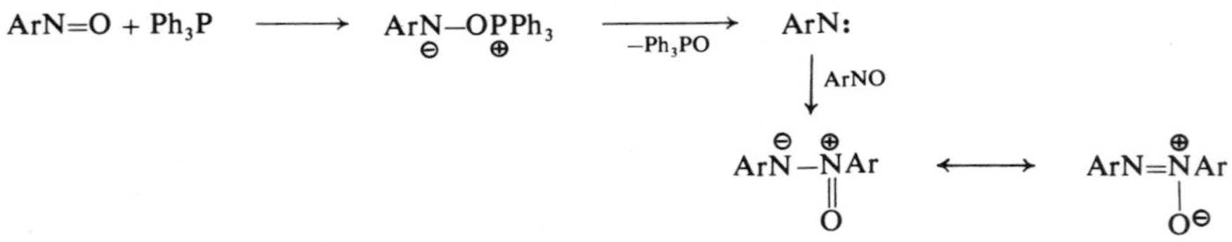

Similarly, aromatic nitroso compounds are deoxygenated. The nitrenes are intercepted to yield azoxyarenes (73). The internal trapping of the nitrene or the dipolar precursor (74) leads to heterocyclic products. Thus, 2-nitroso-biphenyl affords carbazole (75) in good yield. Spirodienyl intermediates are in-

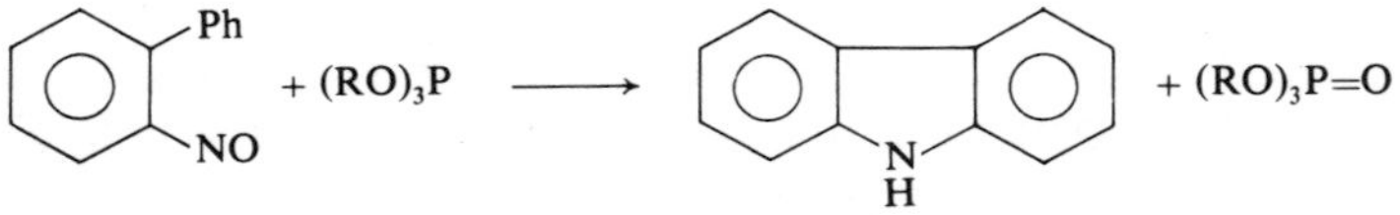

volved in the deoxygenation rearrangement of certain aromatic nitro compounds (76). The phosphite reagent serves the dual function of removing oxygen atoms from the nitro group and effecting O—C bond cleavage of the spiro species.

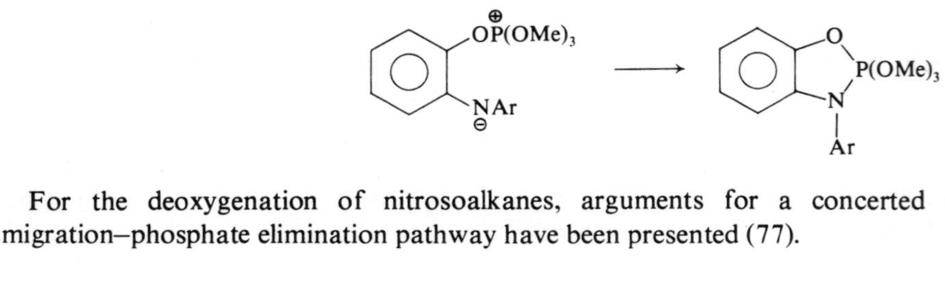

For the deoxygenation of nitrosoalkanes, arguments for a concerted migration—phosphate elimination pathway have been presented (77).

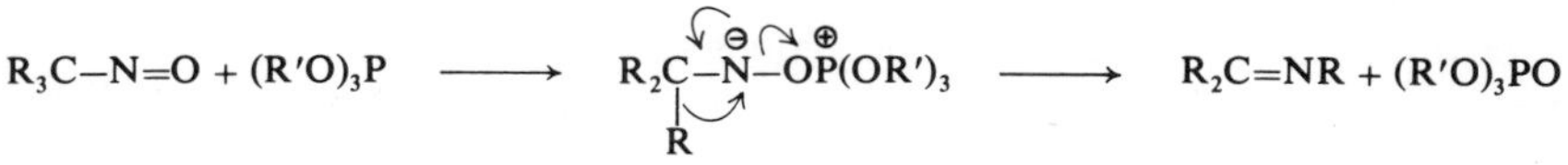

$$R_3C-N=O + (R'O)_3P \longrightarrow R_2C-N-OP(OR')_3 \longrightarrow R_2C=NR + (R'O)_3PO$$
$$\overset{|}{R}$$

Semipolar oxygen may be removed by P(III) derivatives. However, the opposite reactivity order of these reagents to their usual nucleophilicities suggests that a biphilic process (78) is involved in the reaction. The two reactivity

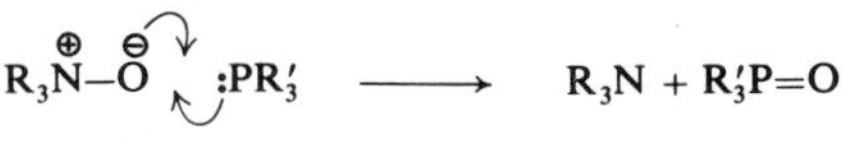

$$R_3\overset{\oplus}{N}-\overset{\ominus}{O} \quad :PR'_3 \longrightarrow R_3N + R'_3P=O$$

sequences for the trivalent phosphorus compound reduction of dimethyl sulfoxide (79) may indicate a change in the relative importance of σ and π donation during a biphilic attack on oxygen.

$$PCl_3 > (PhO)_3P > PhPCl_2 > Ph_2PCl > (MeO)_3P > Ph_3P$$

$$(Me_2N)_3P > n\text{-}Bu_3P > Ph_3P$$

The cleavage of allylic sulfenate esters by phosphites (80) may actually involve attack at sulfur, even though the products indicate otherwise.

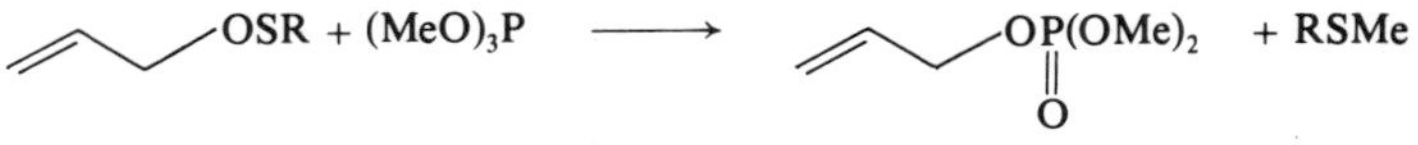

$$\diagup\!\!\!\!\diagup\diagdown\!\diagup OSR + (MeO)_3P \longrightarrow \diagup\!\!\!\!\diagup\diagdown\!\diagup OP(OMe)_2 + RSMe$$
$$\underset{O}{\overset{\|}{}}$$

Phosphines and phosphites are readily oxidized by ozone (81). The 1:1 adducts have pentacovalent phosphorus structures (82). Ozonides are reductively cleaved by triphenylphosphine (83).

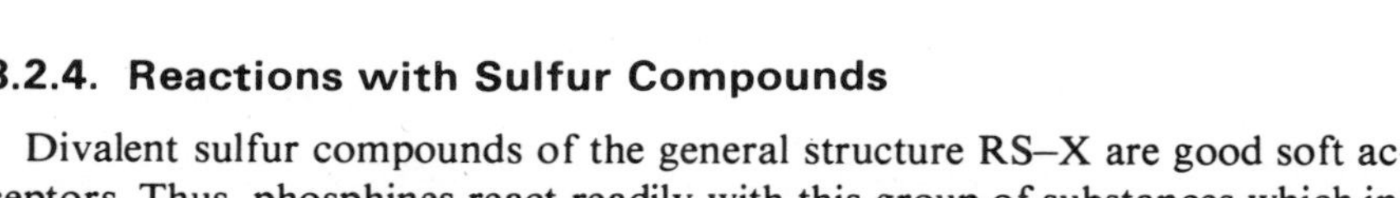

The sulfur atom of sulfinyl (84) and sulfonyl derivatives (85) is reduced to the divalent state on treatment with P(III) compounds.

8.2.4. Reactions with Sulfur Compounds

Divalent sulfur compounds of the general structure RS—X are good soft acceptors. Thus, phosphines react readily with this group of substances which include disulfides (86) and polysulfides (87). Interestingly, phosphines of different softness attack different sites of trisulfides (88) as illustrated in the following equations. The softer phosphines show a higher affinity for the central sulfur atom which is a softer acceptor because it is flanked by two soft atoms.

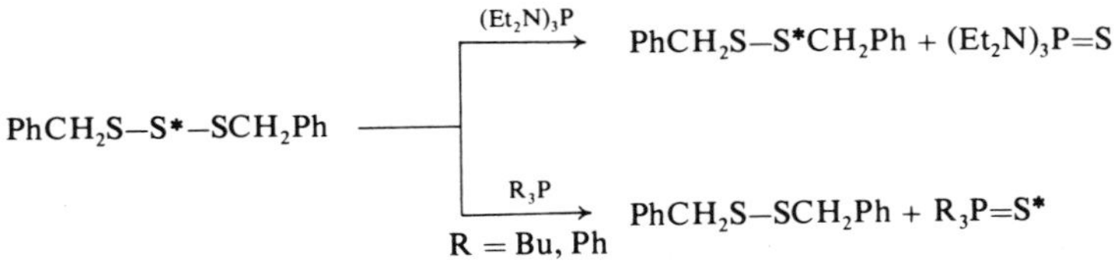

$$\text{PhCH}_2\text{S--S*--SCH}_2\text{Ph} \quad \begin{array}{l} \xrightarrow{\text{(Et}_2\text{N)}_3\text{P}} \quad \text{PhCH}_2\text{S--S*CH}_2\text{Ph} + \text{(Et}_2\text{N)}_3\text{P=S} \\[2em] \xrightarrow[R = \text{Bu, Ph}]{R_3\text{P}} \quad \text{PhCH}_2\text{S--SCH}_2\text{Ph} + R_3\text{P=S*} \end{array}$$

In the presence of water, triphenylphosphine reduces diaryl disulfides to arenethiols (89, 90). An intramolecular version of this double displacement is shown below (91).

$$\text{Ph}_3\text{P} + \text{ArSSAr} \longrightarrow \text{Ph}_3\overset{\oplus}{\text{P}}\text{SAr} \ \overset{\ominus}{\text{SAr}} \ \xrightarrow{\text{H}_2\text{O}} \ 2\text{ArSH} + \text{Ph}_3\text{P=O}$$

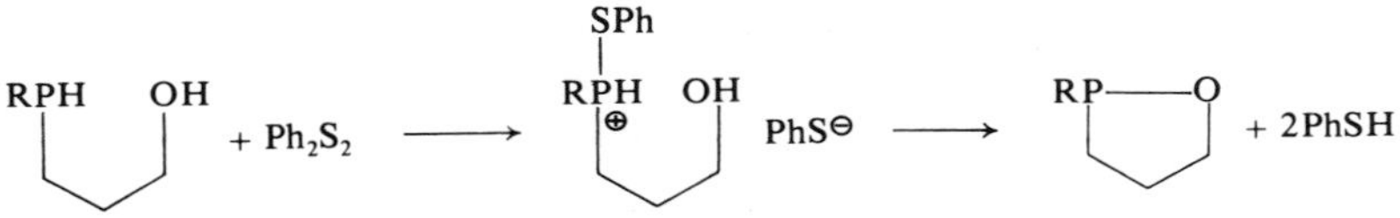

The formation of thiiranes (92) by the following reaction illustrates another example consistent with the HSAB concept.

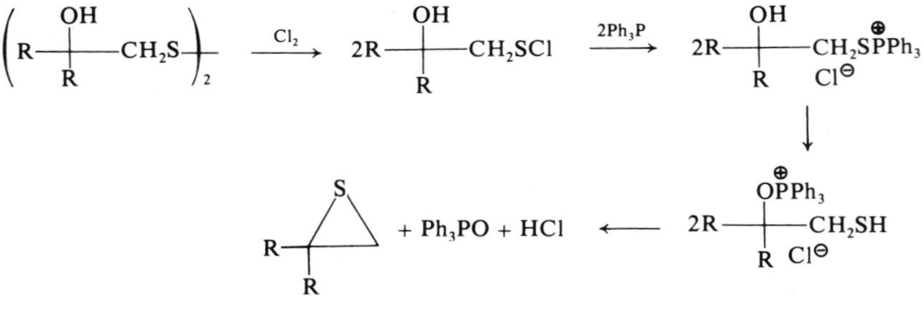

In a peptide synthesis based on redox condensation (93), triphenylphosphine–disulfide adducts are used to activate the carboxyl group for the coupling with amines. The overall transformation consists of sequential soft–soft, hard–hard, and hard–hard interactions.

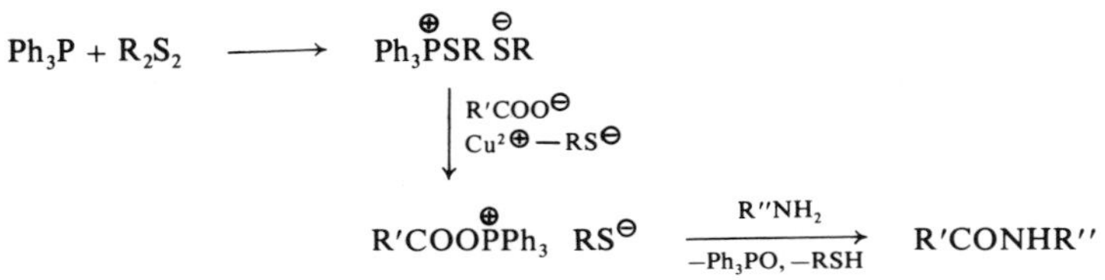

$$\text{Ph}_3\text{P} + R_2\text{S}_2 \longrightarrow \text{Ph}_3\overset{\oplus}{\text{P}}\text{SR} \ \overset{\ominus}{\text{SR}}$$

$$\downarrow \begin{array}{l} R'\text{COO}^{\ominus} \\ \text{Cu}^{2\oplus}\text{--RS}^{\ominus} \end{array}$$

$$R'\text{COO}\overset{\oplus}{\text{P}}\text{Ph}_3 \ \ \text{RS}^{\ominus} \ \xrightarrow[-\text{Ph}_3\text{PO, --RSH}]{R''\text{NH}_2} \ R'\text{CONHR}''$$

Barton and co-workers (94, 95) have developed a useful olefin synthesis, which is particularly applicable to highly hindered alkenes. Hexaethylphosphorous triamide is a key reagent for sulfur abstraction.

Like diacyl peroxides, diacyl disulfides are desulfurized (90) readily by phosphines.

Highly exothermic reactions occur when sulfenyl chlorides are treated with phosphites or phosphines (96). However, the site of initial attack (S or Cl) has not been determined.

Menthyl *P*-methylphosphinite is converted to a phosphonothioate on reaction with sulfenamides (97).

A sulfinate group is displaced from organic thiosulfonates (98) by phosphites, whereas phenyl benzenethiosulfonate undergoes deoxygenation to give diphenyl disulfide by phosphines (99).

The dialkylphosphite ions can displace a sulfite ion from the Bunte salts in a straightforward process (100). Both the donor and the acceptor atoms are soft.

$$(RO)_2PO^\ominus + R'SSO_3^\ominus \longrightarrow (RO)_2\underset{\underset{O}{\|}}{P}SR' + SO_3^{2\ominus}$$

8.2.5. Attack at Nitrogen

Aromatic diazonium salts are reduced by phosphines in protic solvents (101). The crucial steps can be summarized as soft–soft P : N and hard–hard O : P interactions.

$$Ar\overset{\oplus}{N}{\equiv}N \ X^\ominus + R_3P \longrightarrow ArN{=}N{-}\overset{\oplus}{P}R_3 \ X^\ominus$$

$$\downarrow \text{MeOH}$$

$$ArN{=}NH + R_3\overset{\oplus}{P}OMe \ X^\ominus \longrightarrow ArH + N_2 + R_3P{=}O + MeX$$

Organic azides form adducts with phosphines (102) which liberate nitrogen on heating.

$$RN_3 + R_3'P \longrightarrow RN{=}N{-}N{=}PR_3' \overset{\Delta}{\longrightarrow} RN{=}PR_3' + N_2$$

The triphenylphosphine–azodiformate ester combination is a strong dehydrating system in the alkylation of imines by alcohols (103) and the formation of aryl alkyl ethers (104). The synthesis of isonitriles (105) from formamides can be formulated according to the following interactions.

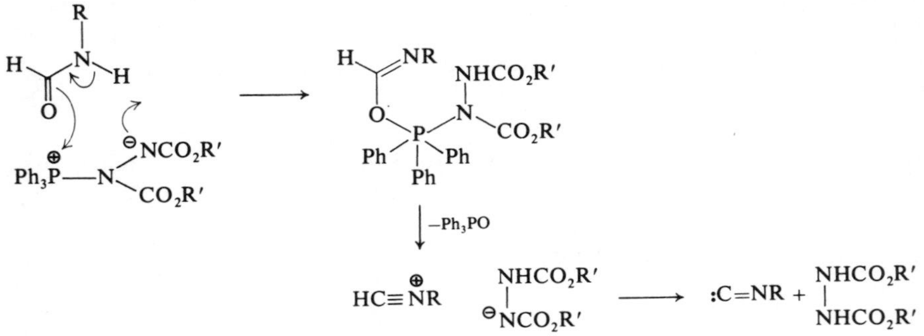

8.3. NUCLEOPHILIC ATTACK ON THE PHOSPHORUS CENTER

8.3.1. Trivalent Phosphorus

Owing to the availability of vacant d orbitals, the trivalent phosphorus center is subject to nucleophilic attack.

The enolate ions react with phosphorochloridites to give enolphosphites (106, 107). However, the dianion of phenylacetic acid is alkylated at the α carbon (108). It appears that these processes are subject to symbiotic control.

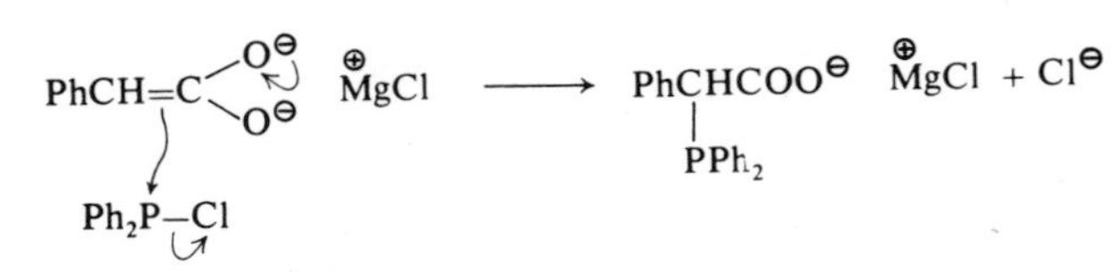

The hydrolysis of chlorodiarylphosphines in the presence of air leads to tetraaryldiphosphine dioxides (109). The P—P bond is formed by electrophilic attack on the chlorophosphine by the conjugate base of the hydrolyzed molecule. This pattern of soft–soft P·P bond making is contrasted to the interaction between diphenylphosphine oxide and chlorodiphenylphosphonate. The phosphoryl chloride has a hard acceptor center, of course.

$$Ar_2PCl \xrightarrow[Et_3N]{H_2O} Ar_2\overset{\ominus}{P}=O \xrightarrow[-Cl^{\ominus}]{Ar_2PCl} \underset{O}{Ar_2P-PAr_2} \xrightarrow{[O]} \underset{O \ \ \ O}{Ar_2P-PAr_2}$$

$$\underset{O}{Ph_2PH} + \underset{O}{Ph_2PCl} \xrightarrow{base} \underset{O}{Ph_2P-O-PPh_2}$$

8.3.2. Tetrahedral Phosphorus

There are two types of P(V) compounds, i.e., the pentacovalent and the ionic phosphonium salts. The most familiar outcome of the attack at a pentacovalent phosphorus atom by nucleophiles is substitutive fragmentation. On the other hand, there are numerous ways for a nucleophile to attack $R_4P^\oplus$ compounds: direct S_N2 displacement, simple addition, and addition–elimination. Many reactions, especially the biphilic processes, involve formation of $R_4P^\oplus$ compounds followed by a counterion-initiated decomposition. The Arbuzov and Perkov reactions are representative.

The phosphine–carbon tetrachloride combination has proven to be a versatile dehydrating agent. Primary amides (110, 111) and aldoximes (112) are converted into nitriles and the substituted ureas are transformed into carbodiimides (113). The reactive species $R_3\overset{\oplus}{P}Cl \ \overset{\ominus}{C}Cl_3$ provides a strongly electrophilic locus for hard donors.

The stereochemistry and reaction modes of dialkoxyphosphonium salts with various nucleophiles (114) can be correlated with the HSAB principle. Epimerization and P–O bond cleavage are associated with the hard bases. The soft bases generally effect C–O bond scission leading to products which retain the configuration of the phosphorus.

Since the phosphonium and pentacovalent phosphorus are hard acceptors, it is not surprising that the secondary phosphine oxides are rapidly oxidized by alkali (115) at room temperature. On the other hand, sodium benzenethiolate effects the oxidative thiolation only very slowly at 70°.

$$\underset{O}{Ph_2PH} + HO^{\ominus} \longrightarrow \underset{\underset{OX}{\overset{H}{|}}}{\overset{\overset{OH}{|}}{Ph_2P-O^{\ominus}}} \longrightarrow \underset{O}{Ph_2POH}$$

$$\underset{O}{Ph_2PH} + PhS^{\ominus} \longrightarrow \underset{O}{Ph_2PSPh}$$

The reductive dimerization (116) of the diphenylphosphonyl chloride and the thiophosphonyl chloride involves hard–hard and soft–soft interactions, respectively, between the anionic intermediates and the starting materials.

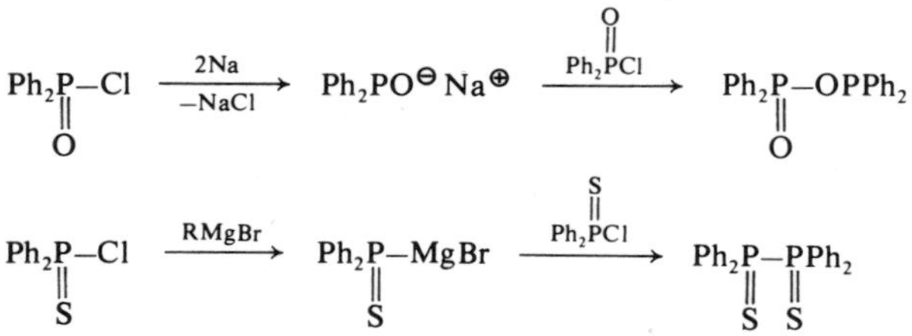

$$Ph_2P\text{--}Cl \xrightarrow[-NaCl]{2Na} Ph_2PO^{\ominus} Na^{\oplus} \xrightarrow{Ph_2PCl} Ph_2P\text{--}OPPh_2$$

(with $\overset{O}{\underset{}{\|}}$ on the starting material, the reagent, and the product)

$$Ph_2P\text{--}Cl \xrightarrow{RMgBr} Ph_2P\text{--}MgBr \xrightarrow{Ph_2PCl} Ph_2P\text{--}PPh_2$$

(with S groups)

The dealkylation of the ambident trialkylphosphates (117, 118) using soft and hard bases follows separate pathways as expected.

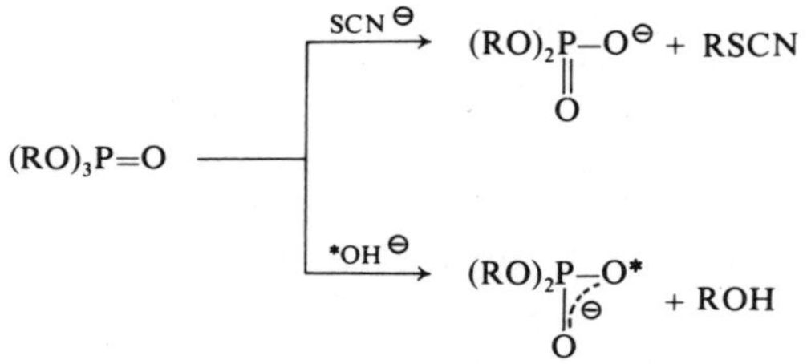

$$(RO)_3P=O \quad \begin{cases} \xrightarrow{SCN^{\ominus}} (RO)_2P\text{--}O^{\ominus} + RSCN \\[2em] \xrightarrow{{}^*OH^{\ominus}} (RO)_2P\text{--}O^* + ROH \end{cases}$$

Acetylcholinesterase is inhibited by organic fluorophosphates through the formation of covalent phosphate esters with the serine residue at the active site of the enzyme. The reactivation of the enzyme can be achieved by treatment with hydroxylamine which acts as a transesterification and not an aminolysis agent (119). The transition state symbiosis may be the determinant for the specificity.

The hydrolysis of dimethyl acetoin phosphate at pH 7.7–8.2 is extremely fast (120), whereas the hydrolysis of methyl acetoin methylphosphonate (121) is 200 times slower. Empirically it can be assumed that the P(V) atom of phosphates is harder than the phosphonates and, accordingly, the phosphates react faster with hard bases.

The α-hydroxyalkylphosphonates undergo facile base-catalyzed mutation to the phosphates (122, 123). The result mirrors the symbiotic stabilization of the transition state and the products.

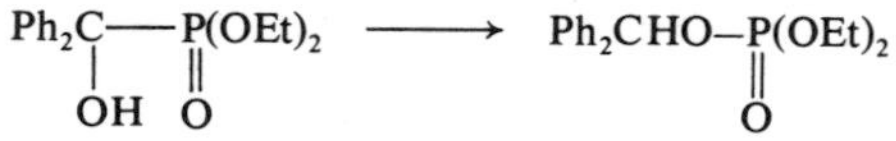

$$Ph_2C\text{--}P(OEt)_2 \longrightarrow Ph_2CHO\text{--}P(OEt)_2$$

(with OH and O substituents on the left, O on the right)

The α-keto phosphonates are converted into similar compounds on treatment with the cyanide ion (124).

$$\underset{\underset{O}{\overset{\parallel}{}}\,\underset{O}{\overset{\parallel}{}}}{MeC-P(OEt)_2} + CN^{\ominus} \longrightarrow \underset{CN\ \ O}{Me-\overset{O^{\ominus}}{\overset{|}{C}}-\overset{\parallel}{P}(OEt)_2} \longrightarrow \underset{CN\quad O}{MeCH-O-\overset{\parallel}{P}(OEt)_2}$$

The mechanism for the thermal rearrangement of bis(diphenylphosphinyl) peroxide to an unsymmetrical anhydride (125) has been elucidated. An intra-molecular hard–hard interaction between oxygen and phosphorus atoms results in a zwitterion intermediate which collapses via migration of a phenyl group from P to O (soft) with a concomitant O–O bond fission.

$$\underset{O^*}{\overset{O^*}{\overset{\parallel}{Ph_2P}}}\diagdown_{O-O}\diagup^{PPh_2} \longrightarrow \underset{\ominus O^*}{Ph_2P}\diagdown_{O}\overset{Ph}{\underset{|}{\overset{\oplus}{P}}-Ph} \longrightarrow \underset{O}{\overset{O^*\quad OPh}{Ph_2P-O-\overset{|}{\underset{\parallel}{P}}-Ph}}$$

The S-alkylphosphorothiolates are hydrolyzed (126) and dialkylphosphoro-hydrazidates are hydrolytically deaminated with the aid of iodine (127). The fragmentation step of the latter reaction is a perfect example of hard/soft cooperativity (Saville's rule).

$$\underset{O}{\overset{O}{\overset{\parallel}{(RO)_2P-NHNH_2}}} \xrightarrow[-2HI]{I_2} (RO)_2-\overset{O}{\overset{\parallel}{P}}-N\!\!=\!\!N\!\!-\!\!H \quad I\!-\!I \longrightarrow \underset{O}{\overset{O}{\overset{\parallel}{(RO)_2P-OH}}} + N_2 + 2HI$$

Phosphoro- and phosphinothiocyanatidates isomerize to the corresponding isothiocyanatidates (128) at room temperature. The lability of the thio-cyanatidates originates from a hard–soft pairing in these compounds.

$$\underset{R'}{\overset{R}{\diagdown}}P\overset{\diagup O}{\diagdown_{SCN}} \longrightarrow \underset{R'}{\overset{R}{\diagdown}}P\overset{\diagup O}{\diagdown_{NCS}}$$

O,S-Diethyl ethylphosphonothioate reacts with water in the presence of silver and fluoride ions (129). The reaction proceeds according to Saville's generalization.

$$X^{\ominus}\diagdown\overset{O}{\underset{Et\ \ OEt}{\overset{\parallel}{P}}}\diagdown_{SEt} + nAg^{\oplus} \longrightarrow \underset{Et}{\overset{O}{\overset{\parallel}{X-P}}}-OEt + EtSAg + (n-1)Ag^{\oplus}$$

$$X = F, OH$$

O,S-Ethylene O-methylphosphorothioate suffers P–S bond cleavage by hydroxide ion (130). The P(V)–S bond is weak because it is a hard–soft combination.

Teichmann and Hilgetag (131) have summarized and discussed the various chemical aspects of the thiophosphoryl group in terms of the HSAB principle. The P=S group may behave as an electrophile (P end) or as a nucleophile (S end).

Optically active phosphine sulfides are desulfurized by $LiAlH_4$ to the parent phosphines with retained configurations (132). The union of the soft hydride ion with the soft sulfur, instead of the hard pentacovalent phosphorus of the thiophosphoryl function, is in keeping with the HSAB principle.

Triarylphosphine sulfides form crystalline adducts with the soft halogens (ICl, IBr, I_2) (133). A facile cleavage of tetraalkyldiphosphine disulfides by halogens (134, 135) is preluded by S–X coordination.

Dithio- (136) and monothiohypophosphates (137) also react exothermically with sulfuryl chloride. However, the sulfur-free analogs must be heated to reflux in order to bring about fission of the P–P bond (138).

$$(RO)_2P-P(OR)_2 + SO_2Cl_2 \longrightarrow (RO)_2P-Cl + (RO)_2P-Cl + SO_2$$
$$\overset{\|}{X}\ \overset{\|}{Z} \qquad\qquad\qquad \overset{\|}{X} \qquad\qquad \overset{\|}{Z}$$

$$X = Z = S$$
$$X = Z = O$$
$$X = S, Z = O$$

The alcoholysis of cyclic phosphorohalidothionates (139) causes inversion of the phosphorus configurations when the halogen is either chlorine or bromine. When the fluoride ion is the leaving group, products with retained configurations are predominant. This latter observation is consequential of symbiotic stabilization of the trigonal-bipyramidal intermediate(s) by fluoride ion such that its expulsion is slower than pseudorotation about the phosphorus center.

The ambient behavior of phosphorus derivatives should be mentioned. Hexamethylphosphoric triamide forms a simple phosphonium salt on alkylation with dimethyl sulfate (140), while it affords a dimeric product with methyl iodide (141). The results fit nicely into the HSAB framework.

$$(Me_2N)_3P=O \xrightarrow{\ Me_2SO_4\ } (Me_2N)_3\overset{\oplus}{P}OMe \quad MeSO_4^{\ominus}$$

$$\Big\downarrow MeI$$

$$(Me_2N)_2\overset{\oplus}{\underset{\overset{\|}{O}}{P}}NMe_3 \quad I^{\ominus} \xrightarrow[-Me_3N]{(Me_2N)_3P=O} (Me_2N)_3\overset{\oplus}{P}-O-\underset{\overset{\|}{O}}{P}(NMe_2)_3$$
$$\qquad\qquad\qquad\qquad\qquad\qquad\qquad\qquad\qquad\qquad I^{\ominus}$$

Phosphabenzenes fail to react with hard acids, yet they readily form σ complexes with soft metal compounds (142). The alkylation of 1,2,4,6-tetraphenylphosphabenzene anion takes place at P, while acylation occurs at the harder C-4 (143).

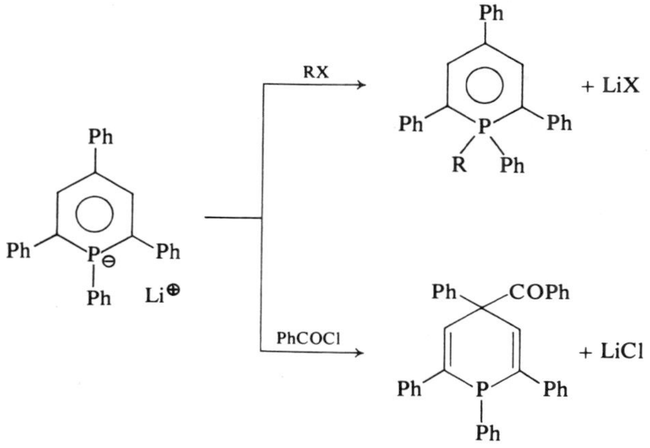

The phosphonothionate and phosphorothiolate ions are alkylated and acylated at different sites (144) as one would anticipate on the basis of the

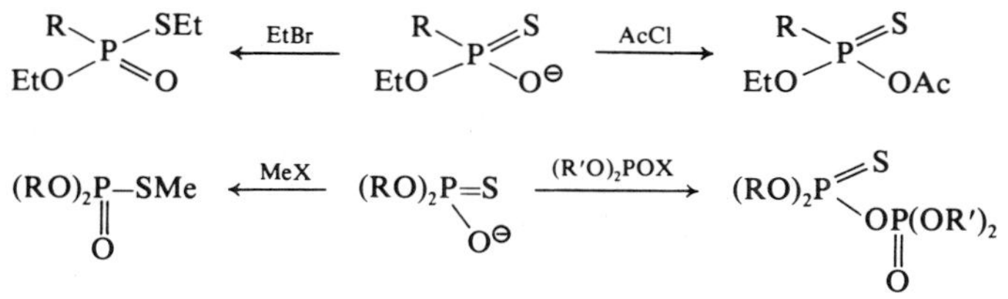

HSAB axiom. However, results contradictory to HSAB prognostication are known in reactions involving ambient ions, e.g.,

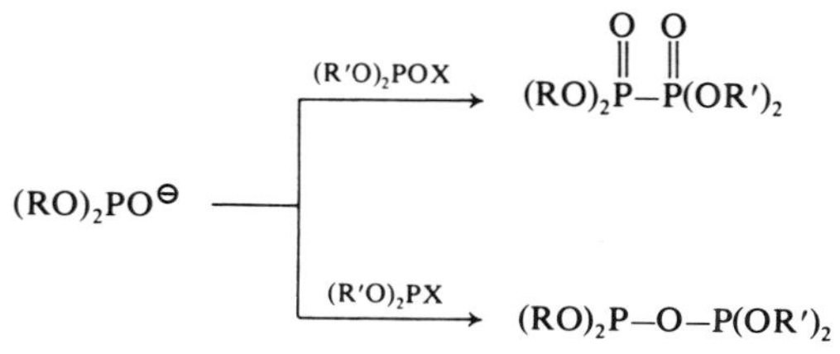

In such cases, the interpretation requires a knowledge of the nature of the transition state (145). Thus, P-phosphorylation is determined by a high P–O bond energy in the transition state which resembles the addition intermediate. A

thermodynamically stable product results. In the second reaction outlined above, the greater charge on oxygen dictates its high reactivity as the transition state is closer to the reactants.

REFERENCES

1. M. I. Kabachnik, *Z. Chem.* **2**, 289 (1961).
2. R. G. Pearson, H. B. Gray, and F. Basolo, *J. Am. Chem. Soc.* **82**, 787 (1960).
3. L. Horner and B. Nippe, *Chem. Ber.* **91**, 67 (1958).
4. F. Ramirez and S. Dershowitz, *J. Am. Chem. Soc.* **78**, 5614 (1956).
5. H. Hoffmann, L. Horner, and G. Hassel, *Chem. Ber.* **91**, 58 (1958).
6. R. F. Hudson and P. A. Chopard, *Helv. Chim. Acta* **46**, 2178 (1963).
7. C. Osuch, J. E. Franz, and F. B. Zienty, *J. Org. Chem.* **29**, 3720 (1964).
8. D. Seyferth and J. M. Burlitch, *J. Org. Chem.* **28**, 2463 (1963).
9. E. M. Richards and J. C. Tebby, *Chem. Commun.* p. 957 (1967).
10. D. Seyferth, S. O. Grim, and T. O. Read, *J. Am. Chem. Soc.* **82**, 1510 (1960).
11. I. J. Borowitz, S. Firstenberg, G. B. Borowitz, and D. Schuessler, *J. Am. Chem. Soc.* **94**, 1623 (1972).
12. A. M. Pudovik and V. Avery'anova, *Zh. Obshch. Khim.* **26**, 1426 (1956).
13. T. Mukaiyama, T. Nagaoka, and S. Fukuyama, *Tetrahedron Lett.* p. 2461 (1967).
14. E. Gaydou, G. Peiffer, and A. Guillemonat, *Tetrahedron Lett.* p. 239 (1971).
15. E. M. Gaydou and J. P. Bianchini, *Chem. Commun.* p. 541 (1975).
16. V. Mark, *J. Am. Chem. Soc.* **85**, 1884 (1963).
17. J. Songstad, *Acta Chem. Scand.* **21**, 1681 (1967).
18. C. Benezra and J.-L. Bravet, *Can. J. Chem.* **53**, 474 (1975).
19. A. E. Arbuzov and E. A. Kasil'nikova, *Izv. Akad. Nauk SSSR, Otd. Khim. Nauk.* p. 30 (1959).
20. K. A. Petrov, A. I. Gavilova, and V. P. Korotkova, *Zh. Obshch. Khim.* **32**, 1978 (1962).
21. J. A. Miller, *Tetrahedron Lett.* p. 4335 (1969).
22. P. A. Chopard, R. F. Hudson, and G. Klopman, *J. Chem. Soc.* p. 1379 (1965).
23. I. J. Borowitz, K. C. Kirby, Jr., P. E. Rusek, and E. W. R. Casper, *J. Org. Chem.* **36**, 88 (1971).
24. I. J. Borowitz, H. Parnes, E. Lord, and K. C. Yee, *J. Am. Chem. Soc.* **94**, 6817 (1973).
25. B. Miller, *J. Org. Chem.* **28**, 345 (1963).
26. B. Miller, *Tetrahedron Lett.* p. 3527 (1964).
27. H. Hoffmann and H. J. Diehr, *Angew. Chem., Int. Ed. Engl.* **3**, 737 (1964).
28. T. Okada and R. Okawara, *Tetrahedron Lett.* p. 2801 (1971).
29. A. J. Speziale and L. R. Smith, *J. Am. Chem. Soc.* **84**, 1868 (1962).
30. H. Hoffmann and H. Förster, *Tetrahedron Lett.* p. 1547 (1963).
31. R. D. Partos and A. J. Speziale, *J. Am. Chem. Soc.* **87**, 5068 (1965).
32. H. G. Viehe and E. Franchimont, *Chem. Ber.* **95**, 319 (1962).
33. H. Hoffmann and H. Förster, *Tetrahedron Lett.* p. 983 (1964).
34. A. Fujii and S. I. Miller, *J. Am. Chem. Soc.* **93**, 3694 (1971).
35. P. Simpson and D. W. Burt, *Tetrahedron Lett.* p. 4799 (1970).
36. D. W. Burt and P. Simpson, *J. Chem. Soc. C*, p. 2872 (1971).
37. W. Hewertson and I. C. Taylor, *Chem. Commun.* p. 119 (1970).
38. H. Hoffmann and H. J. Diehr, *Tetrahedron Lett.* p. 583 (1962).
39. S. Dershowitz and S. Proskauer, *J. Org. Chem.* **26**, 3595 (1961).
40. K. Issleib and D.-W. Müller, *Chem. Ber.* **92**, 3175 (1959).

41. V. Mark, *Tetrahedron Lett.* p. 333 (1961).
42. E. T. McBee, E. P. Wesseler, D. L. Crain, R. Hurnaus, and T. Hodgins, *J. Org. Chem.* **37**, 683 (1972).
43. R. Rabinowitz and R. Marcus, *J. Am. Chem. Soc.* **84**, 1312 (1962).
44. F. Ramirez, N. B. Desai, and N. McKelvie, *J. Am. Chem. Soc.* **84**, 1745 (1962).
45. R. Appel and R. Milker, *Chem. Ber.* **108**, 1783 (1975).
46. P. J. Foley, Jr., *J. Org. Chem.* **34**, 2805 (1969).
47. D. W. Grisley, Jr., J. C. Alm, and C. N. Matthews, *Tetrahedron* **21**, 5 (1965).
48. D. G. Naae and D. J. Burton, *Synth. Commun.* **3**, 197 (1973).
49. D. Seyferth, J. K. Heeren, and S. O. Grim, *J. Org. Chem.* **26**, 4783 (1962).
50. A. K. Tsolis, W. E. McEwen, and C. A. Van der Werf, *Tetrahedron Lett.* p. 3217 (1964).
51. D. B. Denney and R. R. DiLeone, *J. Am. Chem. Soc.* **84**, 4737 (1962).
52. D. B. Denney and J. W. Hanifin, *Tetrahedron Lett.* p. 2177 (1963).
53. J. D. Roberts and M. C. Caserio, "Basic Principles of Organic Chemistry," Benjamin, New York, 1964.
54. L. Horner and W. Jurgeleit, *Justus Liebigs Ann. Chem.* **591**, 138 (1955).
55. M. A. Greenbaum, D. B. Denney, and A. K. Hoffmann, *J. Am. Chem. Soc.* **78**, 2563 (1956).
56. W. Adam and J. W. Diehl, *Chem. Commun.* p. 797 (1972).
57. W. Adam, J. R. Ramirez, and S. C. Tai, *J. Am. Chem. Soc.* **91**, 1254 (1969).
58. W. Adam and A. Rios, *J. Org. Chem.* **36**, 407 (1971).
59. L. Horner and W. Kirmse, *Justus Liebigs Ann. Chem.* **597**, 66 (1955).
60. F. Hrabák and M. Vacek, *Collect. Czech. Chem. Commun.* **30**, 573 (1965).
61. F. Ramirez, *Pure Appl. Chem.* **9**, 337 (1964).
62. D. B. Denney and H. M. Relles, *J. Am. Chem. Soc.* **86**, 3897 (1964).
63. D. B. Denney, H. M. Relles, and A. K. Tsolis, *J. Am. Chem. Soc.* **86**, 4487 (1964).
64. F. Ramirez and S. Dershowitz, *J. Am. Chem. Soc.* **81**, 4338 (1959).
65. T. Reetz, J. F. Powers, and G. R. Graham, *Abstr. 134th Meet., Am. Chem. Soc., Chicago, 1958,* p. 86P (1958).
66. V. A. Kukhtin and K. M. Kirillova, *Dokl. Akad. Nauk SSSR* **140**, 835 (1961).
67. R. C. Cookson and M. J. Nye, *J. Chem. Soc.* p. 2009 (1965).
68. M. J. Gallagher and I. D. Jenkins, *J. Chem. Soc. C* p. 2605 (1969).
69. A. J. Floyd, K. C. Symes, G. I. Fray, G. E. Gymer, and A. W. Oppenheimer, *Tetrahedron Lett.* p. 1735 (1970).
70. J. A. Miller, *Tetrahedron Lett.* p. 3427 (1970).
71. F. Ramirez, H. Yamanaka, and O. H. Basedow, *J. Am. Chem. Soc.* **83**, 173 (1961).
72. T. Mukaiyama, H. Nambu, and M. Okamoto, *J. Org. Chem.* **27**, 3651 (1962).
73. L. Horner and H. Hoffmann, *Angew. Chem.* **68**, 478 (1956).
74. P. K. Brooke, R. B. Herbert, and F. G. Holliman, *Tetrahedron Lett.* p. 761 (1973).
75. P. J. Bunyan and J. I. G. Cadogan, *J. Chem. Soc.* p. 42 (1963).
76. J. I. G. Cadogan, D. S. B. Grace, P. K. K. Lim, and B. S. Tait, *Chem. Commun.* p. 520 (1972).
77. R. A. Abramovitch, J. Court, and E. P. Kyba, *Tetrahedron Lett.* p. 4059 (1972).
78. F. Ramirez and A. M. Aguiar, *Abstr. 134th Meet., Am. Chem. Soc., Chicago, 1958,* 42N.
79. E. H. Amondoo-Neizer, S. K. Ray, R. A. Shaw, and B. C. Smith, *J. Chem. Soc.* p. 4296 (1965).
80. D. A. Evans and G. C. Andrews, *J. Am. Chem. Soc.* **94**, 3672 (1972).
81. L. Horner, H. Schäfer, and W. Ludwig, *Chem. Ber.* **91**, 75 (1958).
82. Q. E. Thompson, *J. Am. Chem. Soc.* **83**, 845 (1961).
83. O. Lorenz and C. R. Parks, *J. Org. Chem.* **30**, 1976 (1965).

84. A. C. Poshkus and J. E. Herweh, *J. Am. Chem. Soc.* **84**, 555 (1962).
85. L. Horner and H. Nickel, *Justus Liebigs Ann. Chem.* **597**, 20 (1955).
86. R. G. Harvey, H. I. Jacobson, and E. V. Jensen, *J. Am. Chem. Soc.* **85**, 1618 (1963).
87. D. N. Harpp, J. G. Gleason, and J. P. Snyder, *J. Am. Chem. Soc.* **90**, 4181 (1968).
88. D. N. Harpp and D. K. Ash, *Chem. Commun.* p. 811 (1970).
89. A. Schönberg, *Ber. Dtsch. Chem. Ges. B* **68**, 163 (1935).
90. A. Schönberg and M. Z. Barakat, *J. Chem. Soc.* p. 892 (1949).
91. M. Grayson and C. E. Farley, *Chem. Commun.* pp. 830, 831 (1967).
92. J. E. Baldwin and D. P. Hesson, *Chem. Commun.* p. 667 (1976).
93. R. Matsueda, H. Maruyama, M. Ueki, and T. Mukaiyama, *Bull. Chem. Soc. Jpn.* **44**, 1373 (1971).
94. D. H. R. Barton and B. J. Willis, *Chem. Commun.* p. 1225 (1970).
95. D. H. R. Barton, E. H. Smith, and B. J. Willis, *Chem. Commun.* p. 1226 (1970).
96. D. C. Morrison, *J. Am. Chem. Soc.* **77**, 181 (1955).
97. H. P. Benschop, D. H. J. M. Platenburg, F. H. Meppelder, and H. L. Boter, *Chem. Commun.* p. 33 (1970).
98. J. Michalski, T. Modro, and J. Wieczorkowski, *J. Chem. Soc.* p. 1665 (1960).
99. J. F. Carson and F. F. Wong, *J. Org. Chem.* **26**, 1467 (1961).
100. K. A. Petrov, N. K. Bliznjuk, and V. A. Savostenok, *Zh. Obshch. Khim.* **31**, 1361 (1961).
101. L. Horner and H. Hoffmann, *Angew. Chem.* **68**, 478 (1956).
102. L. Horner and A. Gross, *Justus Liebigs Ann. Chem.* **591**, 117 (1955).
103. O. Mitsunobu, M. Wada, and T. Sano, *J. Am. Chem. Soc.* **94**, 679 (1972).
104. M. S. Manhas, W. H. Hoffman, B. Lal, and A. K. Bose, *J. Chem. Soc., Perkin Trans. 1*, p. 461 (1975).
105. B. Beijer, E. von Hinrichs, and I. Ugi, *Angew. Chem.* **84**, 957 (1972).
106. I. F. Lutsenko and Z. S. Kraits, *Zh. Obshch. Khim.* **32**, 1663 (1962).
107. M. I. Kabachnik, P. A. Rossiiskaya, M. P. Shabanova, D. M. Paikin, L. F. Efimova, and N. M. Gamper, *Zh. Obshch. Khim.* **30**, 2218 (1960).
108. S. Raines, *Diss. Abstr.* **24**, 2702 (1964).
109. L. D. Quin and H. G. Anderson, *J. Org. Chem.* **31**, 1206 (1966).
110. E. Yamato and S. Sugasawa, *Tetrahedron Lett.* p. 4383 (1970).
111. R. Appel, R. Kleinstück, and K.-D. Ziehn, *Chem. Ber.* **104**, 1030 (1971).
112. R. Appel, R. Kleinstück, and K.-D. Ziehn, *Chem. Ber.* **104**, 2025 (1971).
113. R. Appel, R. Kleinstück, and K.-D. Ziehn, *Chem. Ber.* **104**, 1335 (1971).
114. K. E. DeBruin and S. Chandrasekaren, *J. Am. Chem. Soc.* **95**, 974 (1973).
115. I. G. M. Campbell and I. D. R. Stevens, *Chem. Commun.* p. 505 (1966).
116. T. Emoto, H. Gomi, M. Yoshifuji, R. Okazaki, and N. Inamoto, *Bull. Chem. Soc. Jpn.* **47**, 2449 (1974).
117. R. F. Hudson and D. C. Harper, *J. Chem. Soc.* p. 1356 (1958).
118. J. R. Cox and O. B. Ramsay, *Chem. Rev.* **64**, 343 (1964).
119. I. B. Wilson, *J. Biol. Chem.* **190**, 111 (1951), and later papers.
120. F. Ramirez, B. Hanson, and N. B. Desai, *J. Am. Chem. Soc.* **84**, 4588 (1962).
121. D. S. Frank and D. A. Usher, *J. Am. Chem. Soc.* **89**, 6360 (1967).
122. H. Timmler and I. Kurz, *Chem. Ber.* **104**, 3740 (1971).
123. U. Hasserodt and F. Korte, *Angew. Chem., Int. Ed. Engl.* **2**, 98 (1963).
124. L. A. R. Hall, C. W. Stephens, and J. J. Drysdale, *J. Am. Chem. Soc.* **79**, 1768 (1957).
125. R. L. Dannley, R. L. Waller, R. V. Hoffman, and R. F. Hudson, *Chem. Commun.* p. 1362 (1971).
126. T. Wieland and R. Lambert, *Chem. Ber.* **89**, 2476 (1956).
127. D. M. Brown and N. K. Hamer, *Proc. Chem. Soc., London* p. 212 (1964).

128. A. Łopusiński, J. Michalski, and W. J. Stec, *Angew. Chem., Int. Ed. Engl.* **14**, 108 (1975).
129. B. Saville, *J. Chem. Soc.* p. 4624 (1961).
130. D. C. Gay and N. K. Hamer, *Chem. Commun.* p. 1564 (1970).
131. H. Teichmann and G. Hilgetag, *Angew. Chem., Int. Ed. Engl.* **6**, 1013 (1967).
132. R. Luckenbach, *Tetrahedron Lett.* p. 2177 (1971).
133. W. Tefteller and R. A. Zingaro, *Inorg. Chem.* **5**, 2151 (1966).
134. W. Kuchen, H. Buchwald, K. Strolenberg, and J. Matten, *Justus Liebigs Ann. Chem.* **652**, 28 (1962).
135. L. Maier, *Chem. Ber.* **94**, 3051 (1961).
136. L. Almasi and L. Paskucz, *Chem. Ber.* **96**, 2024 (1963).
137. J. Michalski, W. Stec, and A. Zwierzak, *Bull. Acad. Pol. Sci., Ser. Sci. Chim.* **13**, 677 (1965).
138. J. Michalski and T. Modro, *Chem. Ind. (London)* p. 1570 (1960).
139. M. Mikolajczyk, J. Krzywanski, and B. Ziemnicka, *Tetrahedron Lett.* p. 1607 (1975).
140. A. Schmidpeter and H. Brecht, *Z. Naturforsch., Teil B* **24**, 179 (1969).
141. C. Anselmi, B. Macchia, F. Macchia, and L. Monti, *Chem. Commun.* p. 1152 (1971).
142. P. Jutzi, *Angew. Chem., Int. Ed. Engl.* **14**, 232 (1975), and references cited therein.
143. G. Märkl and A. Merz, *Tetrahedron Lett.* p. 3611 (1968).
144. M. I. Kabachnik, T. A. Mastryukova, N. I. Kurochkin, N. P. Rodionova, and E. M. Popov, *Zh. Obshch. Khim.* **26**, 2228 (1956).
145. R. F. Hudson, *Coord. Chem. Rev.* **1**, 89 (1966).

9

Reactions of Organosulfur Compounds and Other Chalcogenides

9.1. CHEMICAL SELECTIVITY OF ORGANIC CHALCOGENIDES

Group VIA in the Periodic Table contains such important elements as oxygen and sulfur. The divalent oxygen is a hard base, the others are soft.

Acetals are very labile toward aqueous acids, whereas the thioacetals are remarkably stable because of the softness of the sulfur atoms. Hydrolytic cleavage of the latter class of compounds almost invariably involves coordination with soft or borderline acceptors such as Hg(II) (1–3), Tl(III) (4), Ag(I) (5, 6), Cu(II) (7), $R^\oplus$ (8–11), Hal$^\oplus$ (12), and $NH_2^\oplus$ (13). Chloramine T (14) may be considered as a nitrene complex, hence a soft Lewis acid.

Secondary isotope effects (15) and nmr evidence (16) clearly show that a C—O bond scission occurs during the acid hydrolysis of oxathiolanes. Proton-transfer rates for acidic alcohols are several orders of magnitude higher than those for the corresponding thiols (17). These species-specific interactions are in good agreement with the HSAB dictum.

The S—C=O linkage is susceptible to attack by hard bases (18). The inherent reactivity is caused by the soft–hard union. Complex thioesters can be induced to cyclize to macrolides by the catalysis of mercuric salts (19, 20). Heavy metal ion-assisted aminolysis of thioesters (21) is also known.

$$\text{PhCOSR} + \text{PhNH}_2 + \text{Pb}^{2\oplus} \longrightarrow \text{PhCONHPh} + \text{RSPb}^\oplus + \text{H}^\oplus$$

The ease of anhydro sugar formation (22) from alkali treatment of the chalcogen isolog (1) decreases in the order X = Se ≥ S > O.

The major factor governing the glucoside stability toward alkali is considered by this author to be symbiosis at the anomeric center. Thus, the ioniza-

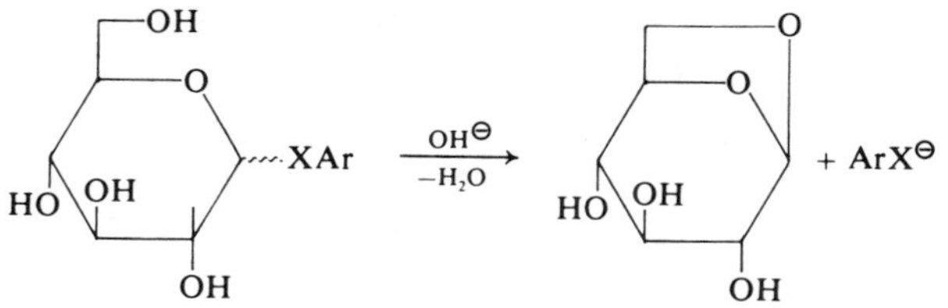

(1)

tion of $ArX^{\ominus}$ is easier when X is softer, as the destabilization of the glucoside is thereby removed.

When submitted to acid hydrolysis (23) these compounds show the reverse order of reactivities: $X = O \gg S > Se$. It is reasonable to assume that under such conditions protonation at X occurs prior to ionization and ring closure. Of course, the harder atom X protonates better.

Ambident chalcogenides generally react according to the HSAB principle. For example, arylsulfenate anions yield (24) methyl sulfoxides on exposure to MeI, and they give methyl arylsulfenates with the harder methylating agents Me_2SO_4 and $MeOSO_2F$.

Phosphorothioate ions are also O,S-ambident. Their reactions with alkyl halides and sulfenyl chlorides occur at the sulfur (25).

Treatment of N,N,N',N'-tetramethylphosphorodiamidic chloride with hydrogen sulfide gives a mixed anhydride (26) indicating that the intermediary phosphorothioate is phosphorylated at the hard oxygen.

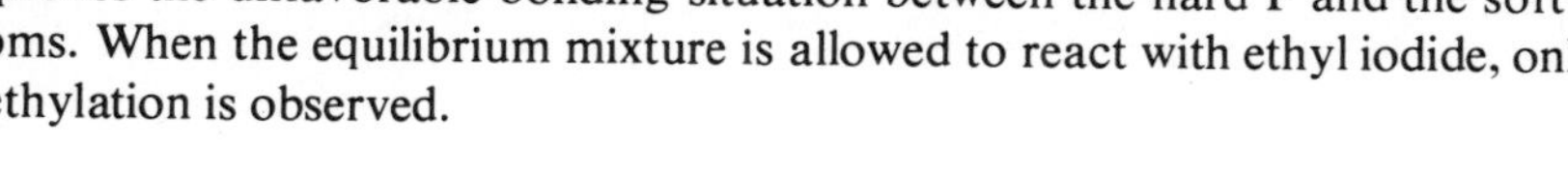

Sodium O,O-dimethylphosphorodithioate is gradually transformed into the O,S-isomer and then into the S,S-isomer on warming (27). The isomerization improves the unfavorable bonding situation between the hard P and the soft S atoms. When the equilibrium mixture is allowed to react with ethyl iodide, only S-ethylation is observed.

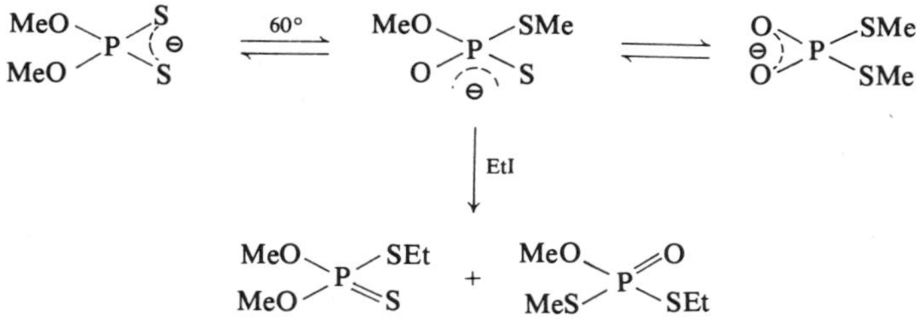

Regiospecific alkylation and acylation of the ambient thiocyanate ion are known to occur at sulfur and nitrogen, respectively.

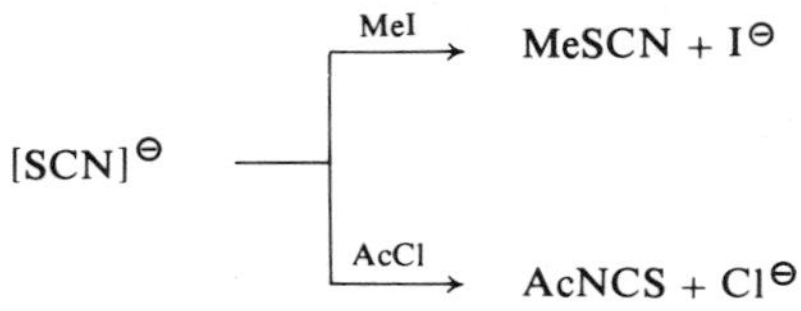

$$[SCN]^{\ominus} \quad \begin{array}{l} \xrightarrow{\text{MeI}} \quad MeSCN + I^{\ominus} \\ \\ \xrightarrow{\text{AcCl}} \quad AcNCS + Cl^{\ominus} \end{array}$$

Mercuric thiocyanate reacts with primary alkyl halides to give alkyl thiocyanates. However, alkyl isothiocyanates become the major products when the reactions are carried out with secondary or tertiary alkyl halides in nonplanar solvents (28). These latter processes are likely to involve ion-pair intermediates in which the harder carbocations bond with the nitrogen end of the complexed thiocyanate counterions.

$$RX + Hg(SCN)_2 \longrightarrow R^{\oplus} XHg(SCN)_2^{\ominus} \longrightarrow RNCS + XHgSCN$$

The deamination of 4-chlorobenzhydrylamine in the presence of thiocyanate ion (29) gives the thiocyanate and isothiocyanate ($k_S:k_N$ 3.8), in addition to solvolytic products. The displacement of alkyl halides by NaSCN in acetonitrile at 50° has also been studied (30). The $k_S:k_N$ ratios (i-PrI, 85; 4-MeOC$_6$H$_4$CH$_2$Br, 206–240; PhCH$_2$Cl, 420–436; PhCH$_2$Br, 835–850; PhCH$_2$I, 1300) vary according to the trend predicted by the HSAB principle.

Following the same pattern, the harder and softer methylating agents, Me$_2$SO$_4$ and MeI, react selectively on N and S termini of thiopurines (31).

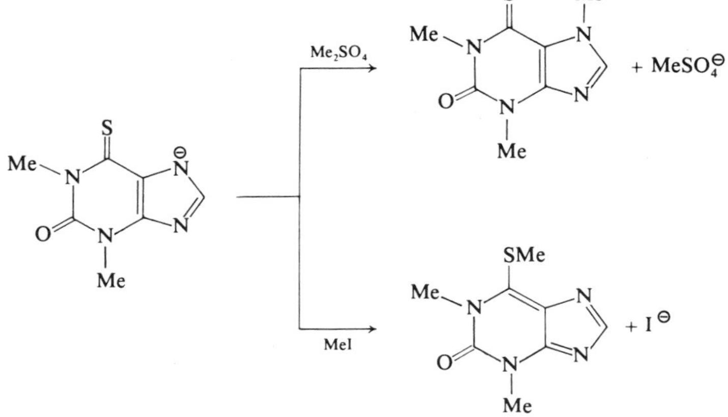

Ethyl 3-acyl-2-thioxo-1,3-oxazolidine-4-carboxylates are acylated (32) and phosphorylated (33) at N and alkylated (34) at S. The various reactions of benzothiazole-2-thiolate ion has been analyzed (35) using the oxibase scale.

Because sulfur is softer than carbon, the rate ratio of 1.7 for the ethylation of the monothiomalonate anion (2) favoring sulfur (36) is most reasonable.

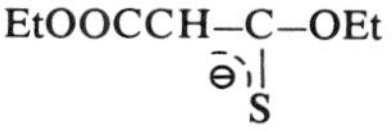

$$\text{EtOOCCH--C--OEt}$$
$$\overset{\ominus}{\underset{\text{S}}{|}}$$

(2)

The treatment of a ketone enolate with carbon disulfide followed by quenching with methyl iodide affords α-di(methylthio)methylene ketones (37). Methylation occurs at both S atoms as contrasted to the alkylation of 2-carbomethoxycyclohexanone enolate which takes place at C.

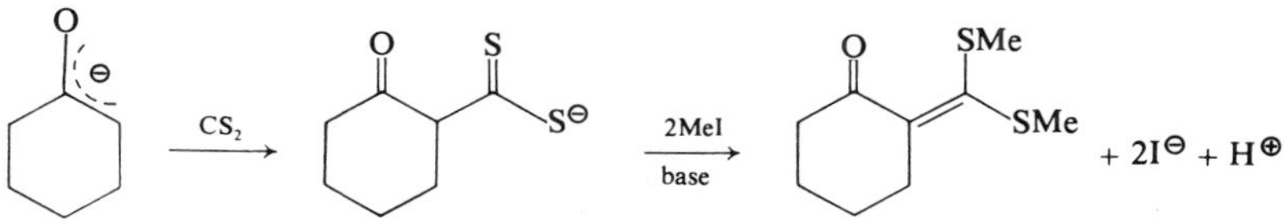

The disparity of O and Se bases is illustrated in the esterification of potassium selenobenzoate (38). Thus, the O-trimethylsilyl esters are produced on reaction with Me_3SiCl, although compounds containing C=Se bonds are very unstable. On the other hand, the salt reacts with trimethylmetal chlorides of more polarizable elements to furnish the seleno esters. The trimethylsilyl ester is sensitive to moisture owing to the presence of a hard Si center. The stannoselenyl and germanoselenyl esters are quite stable toward hard bases.

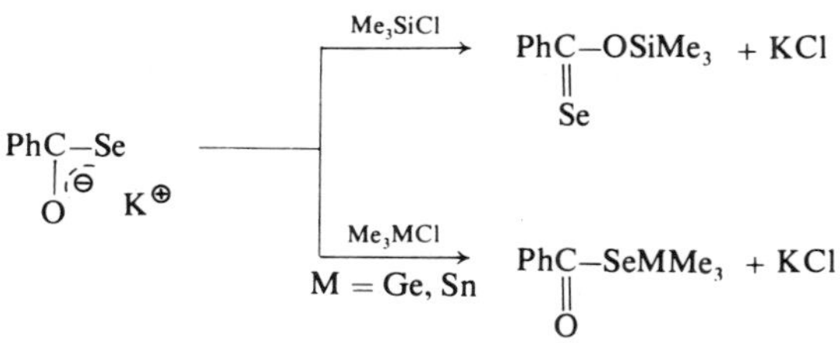

The deselenonation of phosphinoselenoic acids by dimethyl sulfoxide (DMSO) (39) proceeds exothermically, whereas the corresponding reaction with phosphinothioic acids requires heating at 100°. The reactivity of the P=X bond is proportional to the softness disparity of the two adjoining atoms.

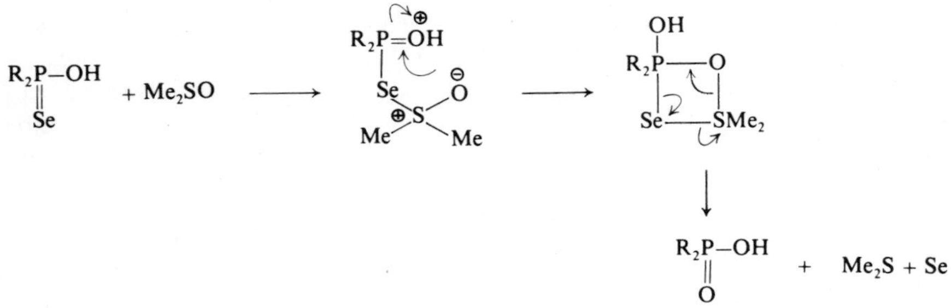

The competitive alkylation and acylation confirm the anticipated relative softness of S and Se. The observed regioselectivity in the reactions of organophosphorus selenothioacids (40) is of interest.

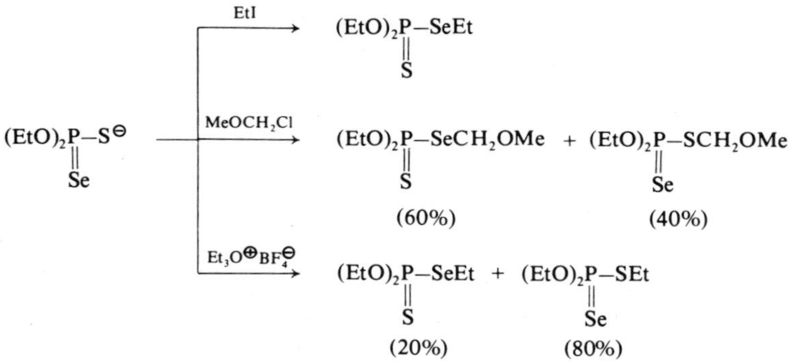

Selenium compounds are typical soft acceptors. The rearrangement of the cyclic phosphoroisoselenocyanatidite (3) on attempted distillation (41) and its loss of Se on treatment with phosphines are cases in point.

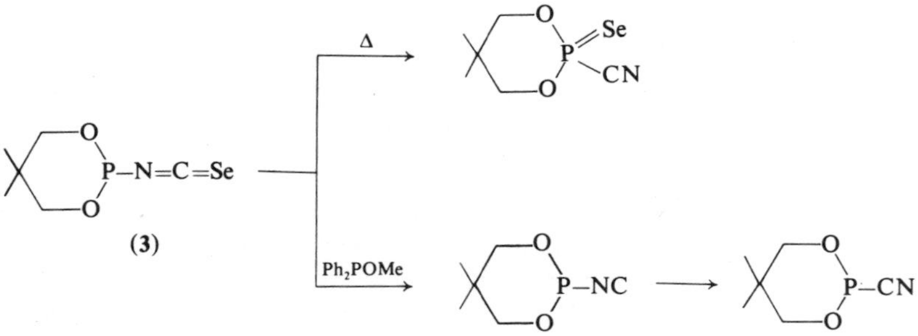

The soft Se atoms of diselenoacetals (42) and selenothioacetals (43) are the sites of attack by organometallic reagents. The comparison with the behavior of dithianes (44) is instructive.

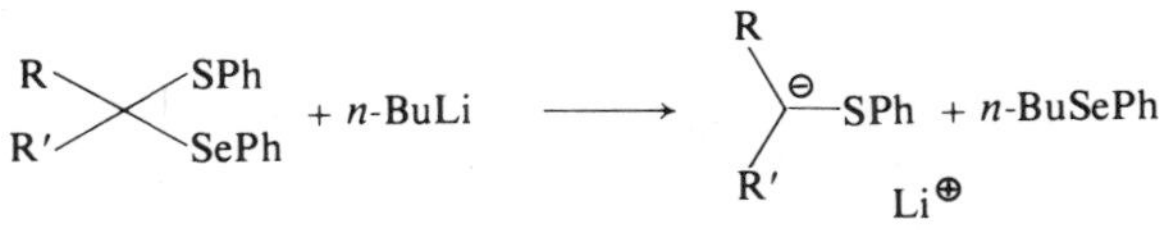

$$\text{(structure)} + n\text{-BuLi} \longrightarrow \text{(structure)} + n\text{-BuSePh}$$

The Se-mediated reactions of alcohols and amines with carbon monoxide (45–47) furnish further evidence for the very soft-acceptor characteristic of elemental selenium. Trapping of the soft carbon monoxide by Se is crucial to the processes.

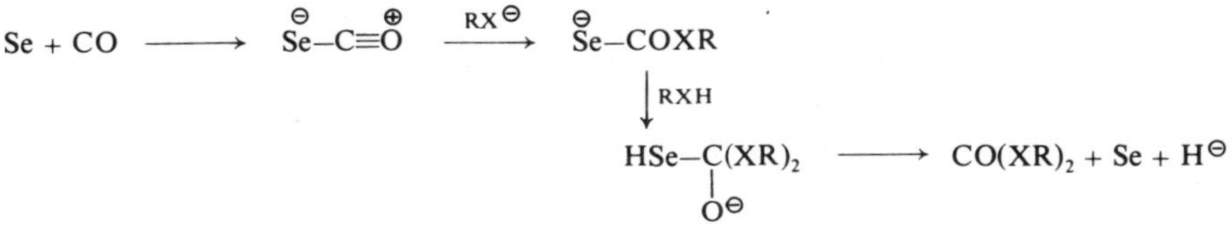

$$\text{Se} + \text{CO} \longrightarrow \overset{\ominus}{\text{Se}}-\overset{\oplus}{\text{C}}{\equiv}\text{O} \xrightarrow{\text{RX}^{\ominus}} \overset{\ominus}{\text{Se}}-\text{COXR}$$

$$\downarrow \text{RXH}$$

$$\text{HSe}-\underset{\underset{\text{O}^{\ominus}}{|}}{\text{C}}(\text{XR})_2 \longrightarrow \text{CO(XR)}_2 + \text{Se} + \text{H}^{\ominus}$$

9.2. DIVALENT SULFUR COMPOUNDS AS SOFT DONORS

Divalent sulfur compounds are highly nucleophilic. However, the activated halogens of α-haloketones are not displaced by thiols or thiolates, owing to the fact that the reagents are soft and the substrates are hardened. Instead, the substrates are reduced (48). Similarly, dimethyl sulfide as such fails to generate α-ketosulfonium salts (49, 50). Only under forcing conditions does Me_2S react with α-bromo-γ-butyrolactone to give α-methylthio-γ-butyrolactone (51), presumably via a biphilic mechanism. In the absence of a proton source the carbanion–bromosulfonium ion pair intermediate resulting from attack on bromine by the sulfur atom can only collapse via C—S bond formation and elimination of methyl bromide.

The conversion of 2-chloro-1,3-diketones to 2-alkylthio-1,3-diketones (52) and the first step of a 1,3-diketone synthesis (53) shown below consist of a biphilic process.

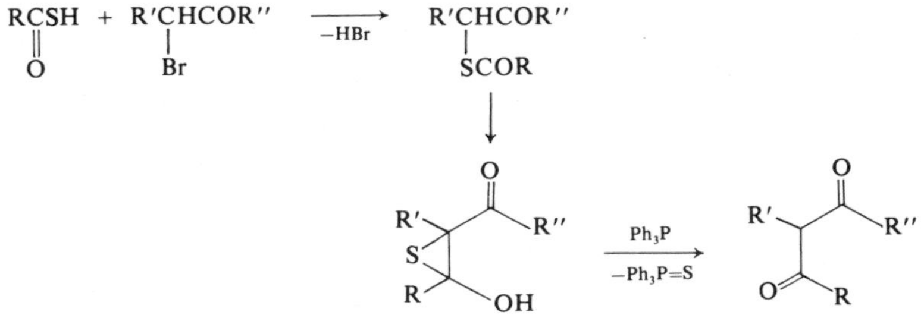

$$\underset{\underset{\text{O}}{\|}}{\text{RCSH}} + \underset{\underset{\text{Br}}{|}}{\text{R'CHCOR''}} \xrightarrow{-\text{HBr}} \underset{\underset{\text{SCOR}}{|}}{\text{R'CHCOR''}}$$

The α-(alkylthio)ketones are dehydrogenated by phenylsulfenyl chloride (54). In this case chlorine transfer between the sulfur atoms of the reactants could be the initial and crucial step. There is little evidence asserting either a biphilic course or direct displacement at S during reaction of thiols with sulfenyl chlorides and acylsulfenyl chlorides (55). Both are soft interactions.

$$ArSH + AcSCl \longrightarrow ArSSAc + HCl$$

The direct conversion of thiols into chlorides has been effected by a soft base-induced fragmentation (56) of the acyl chloride (**4**). The formation of (**4**) from thiols also involves a specific soft–soft interaction.

$$RSH + ClSCOCl \xrightarrow[-HCl]{} \underset{(\mathbf{4})}{RSSCOCl} \xrightarrow[-COS]{Ph_3P} RSPPh_3 \overset{\oplus}{} Cl^{\ominus} \longrightarrow RCl + Ph_3P{=}S$$

It has been demonstrated that halogenation of sulfides proceeds via the Pummerer rearrangement. Isotope, solvent, and reagent effects have been interpreted in terms of the HSAB concept (57).

Both sulfur and olefin linkages are soft bases. The superiority of sulfur in that role is shown by the selective peracid oxidation of dihydrothiophene to the unsaturated sulfone (58).

The rate of ionic decomposition of diacyl peroxides by sulfides (59) are (4-$O_2NC_6H_4CO)_2O_2 > (PhCO)_2O_2 > Ac_2O_2$. The fact that the softer sulfides are more effective is consistent with an initial displacement at the peroxy center by the S bases.

Dimethyl sulfide is a convenient reagent for cleaving ozonides (60) to the carbonyl compounds. The semipolar S→O, S→N, N→O, and N→N bonds are severed by phosphorothiolothionic acids (61) which utilize their sulfur atoms as nucleophiles.

$$R_2SO + (EtO)_2P{-}SH \longrightarrow R_2\overset{\oplus}{S}{-}OH \longrightarrow R_2S + (EtO)_2P{-}SOH$$

The insertion to the O–O bond of dioxetanes by dialkoxysulfides constitutes a novel access to tetralkoxysulfuranes (62). Attack on oxygen and the subsequent charge neutralization steps are favorable on the HSAB basis.

$$(RO)_2S + \quad \longrightarrow$$

The ease of intercepting carbenes by divalent sulfur compounds attests to the soft nature of the S atom. The inhibition of dichlorocarbene addition to cyclohexene (63) by sulfides further demonstrates that sulfides are softer than olefins.

Although coordination of :CCl$_2$ with a dioxolane function was suggested as responsible for the selective cyclopropanation of a diene (64), negative evidence (65) has been accumulated to refute the hard–soft O:C interaction during the reaction. Definitive influence by a dithioacetal group on dichloro-carbene addition to the π bond is indicated (66).

Dichlorocarbene inserts into allyl sulfides (67) via the Stevens rearrange-ment of the ylide intermediates. Internal competition for a carbene is always won by sulfur atom over olefins (68, 69), even though attack on sulfur would give rise to strained molecules.

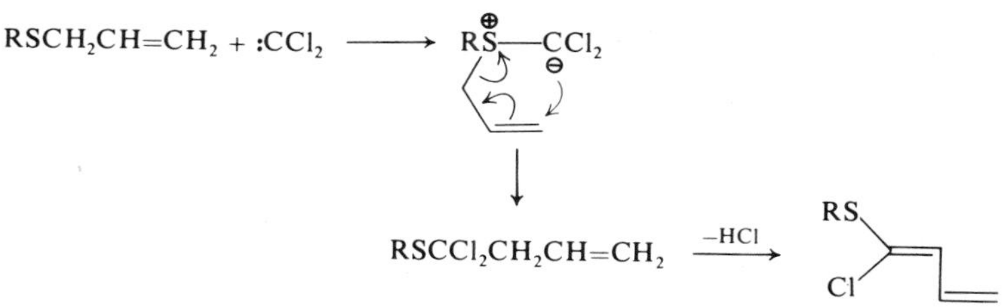

The stereospecific desulfurization of thiiranes may be effected by carbenes (70). Epoxides are deoxygenated with much lower efficiency (71).

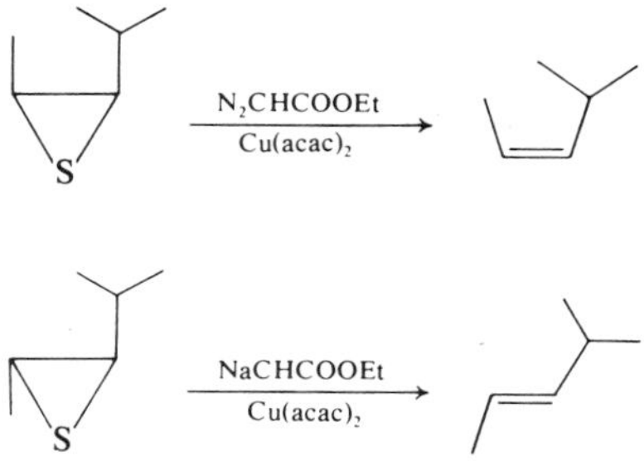

Disulfides are also very reactive toward carbenes (72, 73). The elimination of isobutene has been observed when di-*tert*-butyl disulfide is treated with :CCl$_2$.

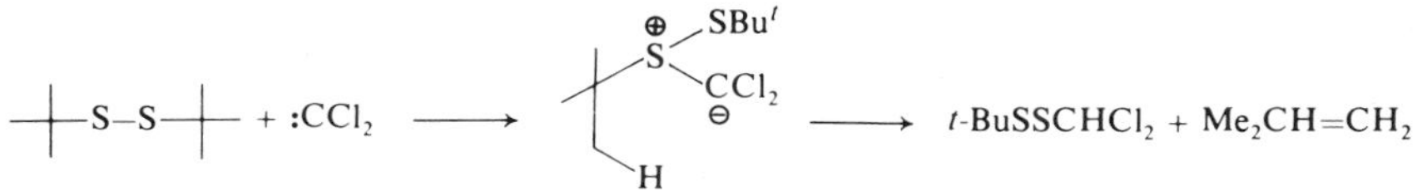

Diethyl disulfide is inserted by phenylphosphene (74), apparently involving a biphilic reaction.

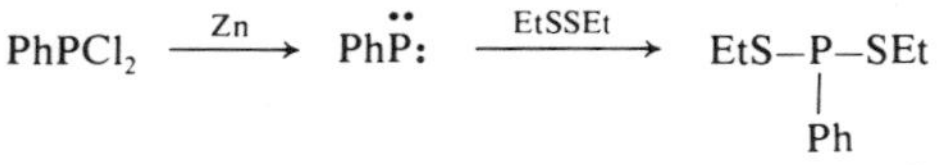

$$PhPCl_2 \xrightarrow{Zn} Ph\overset{..}{P}: \xrightarrow{EtSSEt} \underset{\underset{Ph}{|}}{EtS-P-SEt}$$

The mechanism of sulfo(IV)diimide formation (75) during the reaction of an aminosulfenyl chloride with trimethylsilyl azide can be depicted by nitrene interception and the subsequent elimination of chlorotrimethylsilane from the adduct.

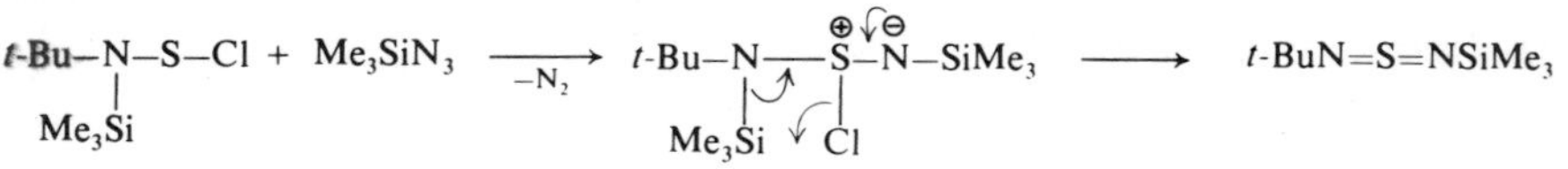

$$\underset{\underset{Me_3Si}{|}}{t\text{-Bu}-N-S-Cl} + Me_3SiN_3 \xrightarrow{-N_2} \quad t\text{-Bu}-N-\overset{\oplus}{S}-\overset{\ominus}{N}-SiMe_3 \longrightarrow t\text{-BuN}=S=NSiMe_3$$

In the total synthesis of the antibiotic cephalosporin C, Woodward et al. (76, 77) introduced a nitrogen function to the methylene group alpha to a sulfur atom in (5) by heating with dimethyl azodiformate. It has been adduced that the unusual process involves an initial addition of the sulfur atom across the azo linkage in a typical soft–soft interaction.

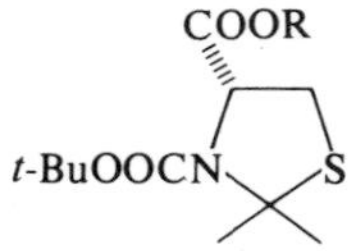

(5)

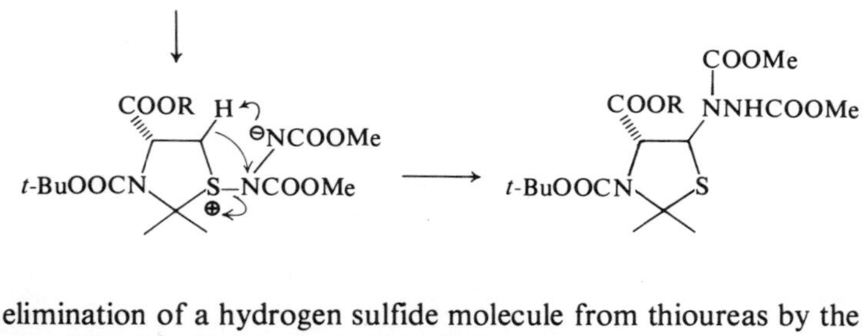

The elimination of a hydrogen sulfide molecule from thioureas by the use of diethyl azodiformate (78) is also initiated by the same S–N bond formation.

The decomposition of diazo compounds in the presence of 1,2-dithiole-3-thiones leads to 3-methylene-1,2-dithiole derivatives (79). The sulfur atom of the thione group acts as a nucleophile toward the incipient carbenic center from the diazo precursors.

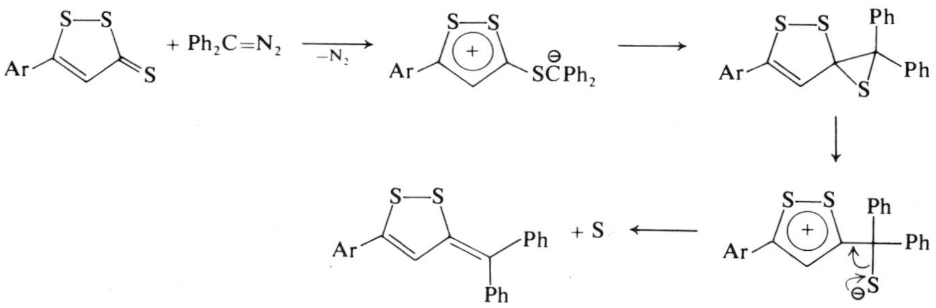

The reactivity of the thiocarbonyl group toward alkyl halides has been evaluated by the CNDO method (80). A much higher energy transition state for the displacement of fluoride ion than of chloride ion is indicated. The demarcation is in keeping with the significantly different softness of the two halogens.

Davis *et al.* (80a) have determined the oxibase parameter (H/E) of some thioanions: $CS_3^{2\ominus}$, 4.27; $EtOCS_2^\ominus$, 1.43; $EtSCS_2^\ominus$, 1.42; $MeCS_2^\ominus$, 1.97; $NCS^\ominus$, 0.55. Thus, comparison can be made of their softness and nucleophilicity.

3-Chloroadamantan-1-ol undergoes fragmentative dehydrochlorination more than 15 times faster than the corresponding thiol (81). The larger frangomeric effect of the former is due to the greater tendency to form a C=O bond rather than the C=S bond.

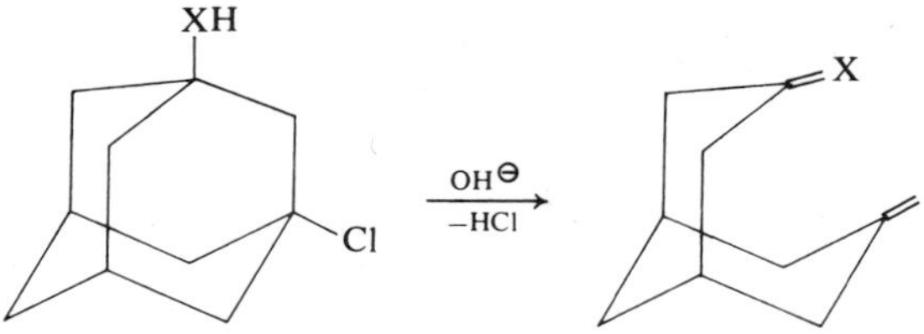

9.3. DIVALENT SULFUR COMPOUNDS AS SOFT ACCEPTORS

The soft donor property of S atom is manifested in sulfur extrusion from thiiranes by carbenes and other soft Lewis acids such as iodine (82), and iodomethane (83); thiiranes assume the role of acceptors when reacting with phosphines (84, 85). Oxiranes, on the contrary, do not react with phosphines except under drastic conditions whereby a mixture of isomeric alkenes are produced.

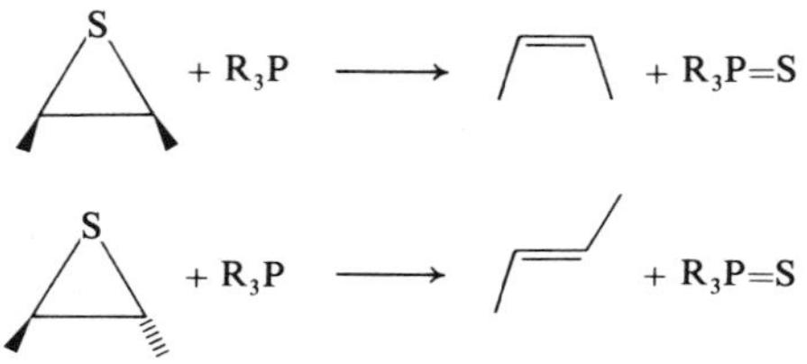

Dewar thiophenes are thiiranes fused to cyclobutene rings. While the reversion of the tetrakis(trifluoromethyl) derivative to the normal thiophene structure requires a high temperature ($\tau_{1/2}[160°$, $C_6H_6]$ 5.1 hr), the transformation can be achieved at room temperature by phosphines (86) such as Ph_3P. A 1:1 adduct can be isolated when Ph_2PCl is added to the Dewar thiophene. However, the harder $PhPCl_2$ and PCl_3 and pentacovalent phosphorus compounds are totally ineffective for the isomerization.

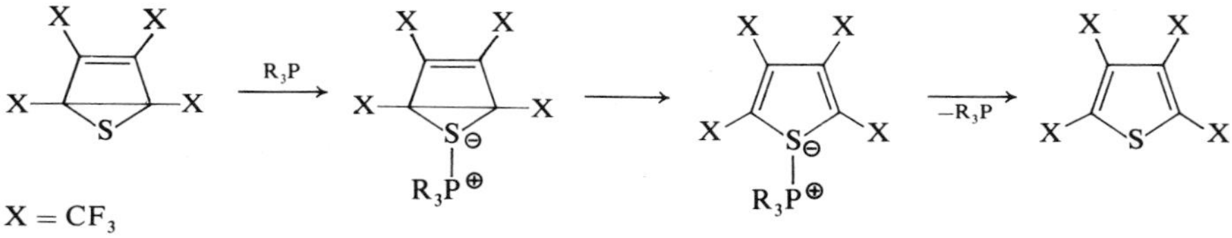

$X = CF_3$

Organolithium reagents elicit fragmentation of thiiranes (87). Neureiter and Bordwell (85) showed that 2-butene episulfides undergo stereospecific desulfuration. Lithium aluminum hydride may also be used (88).

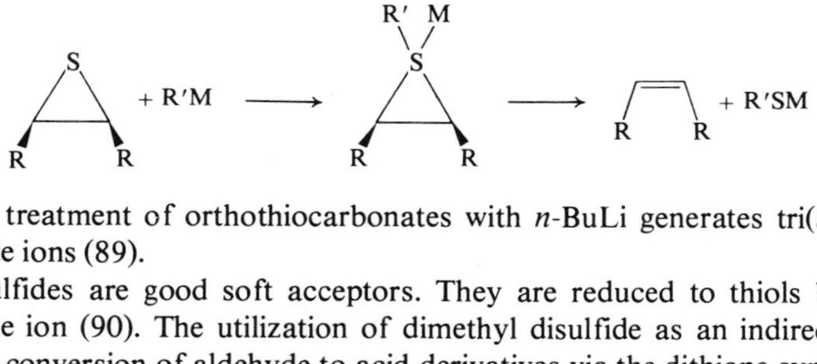

The treatment of orthothiocarbonates with n-BuLi generates tri(alkylthio)-methide ions (89).

Disulfides are good soft acceptors. They are reduced to thiols by hydroselenide ion (90). The utilization of dimethyl disulfide as an indirect oxidant during conversion of aldehyde to acid derivatives via the dithiane synthesis has been reported (91). The alkylation of the lithiodithiane derived from cinnamaldehyde occurs exclusively at the heterocyclic carbon. This may be indicative of symbiotic stabilization of the transition state and the product.

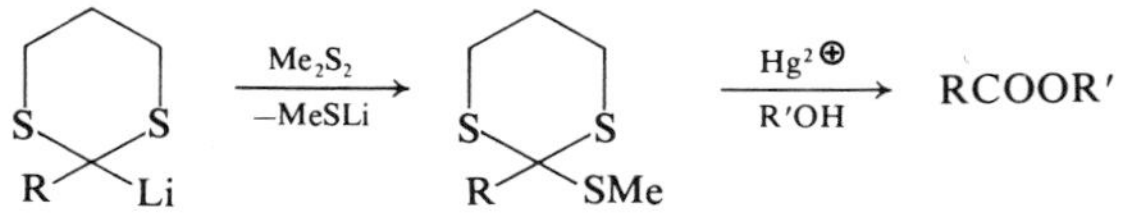

The S–S linkage of sulfenyl thiocyanates, thiosulfates (92), and Bunte salts (93) are also susceptible to cleavage by soft bases.

The displacement on the trithionate ion $S(SO_3)_2^{2\ominus}$ takes place at the central divalent sulfur (94) via a transition state resembling that of the $S_N2(C)$ reaction. The increasing effectiveness of the bases $CN^\ominus < Ph_3P < PhS^\ominus < EtS^\ominus$ has been established.

Mixed disulfides can be synthesized by the reaction of alkyldithioformic esters with thiols (95) through a soft–soft interaction. However, attempts to extend the reaction to prepare sulfenate esters using alkoxides as nucleophiles led to dialkyl trisulfides (96) instead. If a HSAB analysis was performed prior to experiments, the authors would not have been surprised by the results. The hard alkoxide ions would of course elect to attack the ester carbonyl.

$$RSS-COOMe + t\text{-}BuO^\ominus \xrightarrow[-t\text{-BuOCOOMe}]{} RSS^\ominus \xrightarrow{RSS-COOMe} RSSSR + {}^\ominus SCOOMe$$

Heterosubstituted sulfides are excellent sulfenylating agents for soft bases owing to the fact that most of the hetero atoms (e.g., O, N) are hard which favors both the dissolution of the soft–hard combination and the formation of new soft–soft bonds. Thus, the N–S reagents (97–101) are particularly popular for transfer of the thio function.

A synthesis of primary amines has been developed by exploiting the activating and protective ability of PhS groups and their ready removal from amino nitrogen (102a) with soft bases.

$$(PhS)_2NH \xrightarrow[-HX]{RX} (PhS)_2NR \xrightarrow{2PhSH} RNH_2 + 2Ph_2S_2$$

Mixed disulfides are obtained from a sequential treatment of azodiformate esters with two different thiols (102b). An N–S sulfenyl transfer reagent is formed *in situ*.

$$RSH + \begin{matrix} NCO_2Et \\ \| \\ NCO_2Et \end{matrix} \longrightarrow \begin{matrix} RS-NCO_2Et \\ | \\ NHCO_2Et \end{matrix} \xrightarrow{R'SH} RSSR' + \begin{matrix} NHCO_2Et \\ | \\ NHCO_2Et \end{matrix}$$

The sulfur abstraction from *N*-alkylthiophthalimides (103) may be accomplished using a soft nucleophile such as $(Me_2N)_3P$. A biphilic reaction occurs.

3-Chloro-1,2-benzisothiazoles undergo normal substitution at C-3 with ethoxide ion and amines (104). The soft bases attack at S causing rupture of the heterocycle (105).

$Z = CN, n\text{-Bu}$

Since the thione group is much softer than a ketone, diazomethane adds to a thione selectively in the presence of a carbonyl (106). Significantly, the reaction with the thione linkage occurs initially at S.

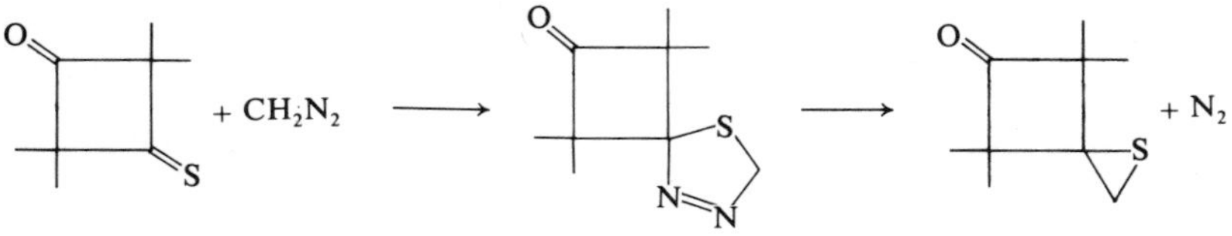

The thione group of β-ketothiones ($\underset{O}{R\overset{\|}{C}}CH_2\underset{S}{\overset{\|}{C}}R$) is more reactive toward (soft) bases (107).

When they react with diazomethane, aromatic thiones are dimerized with incorporation of a methylene group (108). Benzophenone is recalcitrant to this treatment. Hexafluorothioacetone participates in a similar process, but the

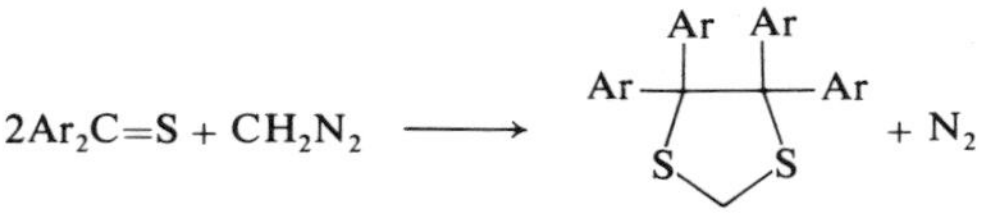

$$2Ar_2C{=}S + CH_2N_2 \longrightarrow$$

product has an unsymmetrical structure (109). Soft nucleophiles bond to the thionic S of hexafluorothioacetone with great ease.

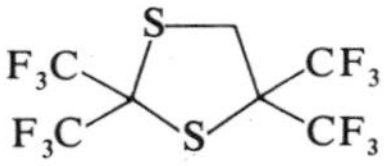

Phosphites can abstract the sulfur from the thione group to generate a carbene. Trithiocarbonates are similarly desulfurized (110) and this reaction forms the basis of the Corey–Winter olefin synthesis (111).

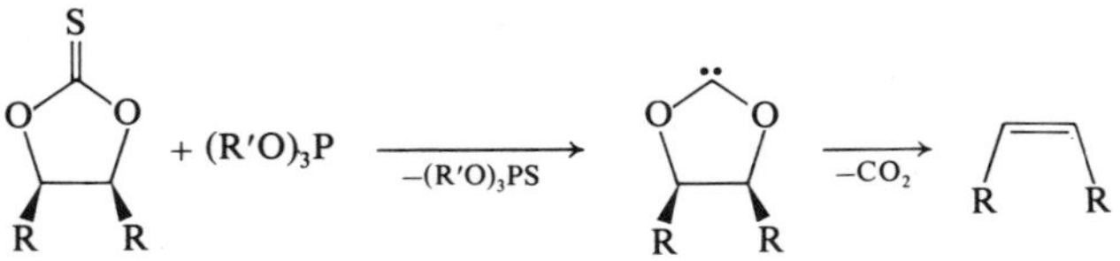

Cyclohexanthione–trialkylphosphite adducts do not decompose into carbene (112) as the latter is not stabilized.

Organometallic agents attack the thionic S leading to a reductive alkylation (113–115). Although it has been demonstrated that the Grignard reaction proceeds via free radical intermediates (116), it is certainly subject to orbital control and obeys the HSAB principle. Hudson (117) has discussed the behavior of such interacting radicals.

$$Ar_2C=S + RM \longrightarrow Ar_2\overset{\ominus}{C}-SR \; M^{\oplus} \longrightarrow Ar_2CHSR$$

$$\downarrow Ar_2C=S$$

$$Ar_2\overset{\frown}{C}-SR \quad M^{\oplus} \longrightarrow \quad Ar_2C\overset{}{\underset{S}{\diagdown\diagup}}CAr_2 \; + RSM$$

$$\underset{\ominus}{\overset{\parallel}{|}}$$

$$S-\overset{}{C}Ar_2$$

The extent of the reaction can be correlated with the softness of the organo-metallic reagents. Thus, the softer MeMgI gives 55% of the adduct with thio-benzophenone, whereas only a 23% yield of the same product is obtained from the corresponding MeMgBr reagent. The harder MeLi attacks the carbon end of the thione moiety as well (118). The dependence of the yield of thiophilic product on the softness of the Grignard reagents is also observed during their reaction with dithiocarboxylic esters (119).

Interestingly, perthiophosphonic anhydrides furnish dithiophosphinic acids (120) on treatment with Grignard reagents. Perhaps the polysulfurated phosphorus atom has become so soft as to be compatible with the attacking species.

$$R-\overset{\overset{\textstyle S}{\parallel}}{P}\overset{S}{\underset{S}{\diagup\diagdown}}\overset{}{\underset{\underset{\textstyle S}{\parallel}}{P}}-R + 2R'MgX \longrightarrow 2RR'P-SMgX \atop \underset{S}{\overset{\parallel}{}}$$

9.4. CHEMISTRY OF THE SULFINYL GROUP

Sulfoxides are ambident bases which can be alkylated at either S or O. Dimethyl sulfoxide gives $Me_2S^{\oplus}OMe \; BsO^{\ominus}$ and $Me_3S^{\oplus}{\rightarrow}O \; I^{\ominus}$ from reaction with methyl brosylate (121) and methyl iodide (122), respectively. The salt for-mation is an S_N2 process whose transition state can be viewed as an acid–base complex, thus the kinetically controlled O-methylation transition state is stabilized symbiotically by the incoming SO and the departing BsO groups. The S-methylation with methyl iodide is dictated by the softness of the alkylator which prefers bonding to the softer S terminus of the sulfinyl group.

Alkyl halides are oxidized to carbonyl compounds by dimethyl sulfoxide (DMSO) at elevated temperatures (123). The drastic conditions can be ob-viated by introducing silver ion into the mixture (124–126). Alternatively, the halides may be converted to the corresponding tosylates (127) which are more susceptible to attack by DMSO. α-Bromoketones (128) and esters (129) yield the dicarbonyl compounds on standing with DMSO.

The HSAB principle provides a sound rationalization for these results. In alkyl halides, the halogen-bound carbon is a soft acceptor and is therefore reluctant to interact with the hard oxy base of the sulfoxide function. By coordination of halides with $Ag^{\oplus}$, the carbon is hardened to a considerable extent. The α carbon of bromoketones and esters is harder and shows greater reactivity toward sulfoxides. The fact that the harder tosylates are better substrates supports this line of reasoning.

Smooth oxidative fission of epoxides to give α-ketols by DMSO is also noted (130). A catalytic amount of boron trifluoride facilitates the transformation by complexing the epoxy oxygen atom, thereby hardening the C termini of the epoxide.

The Moffatt oxidation (131, 132) and its numerous modifications (133) involve activation of DMSO by removing the excess negative charge on O, thus rendering the S atom very hard (as sulfonium center) and receptive to hard bases such as alcohols.

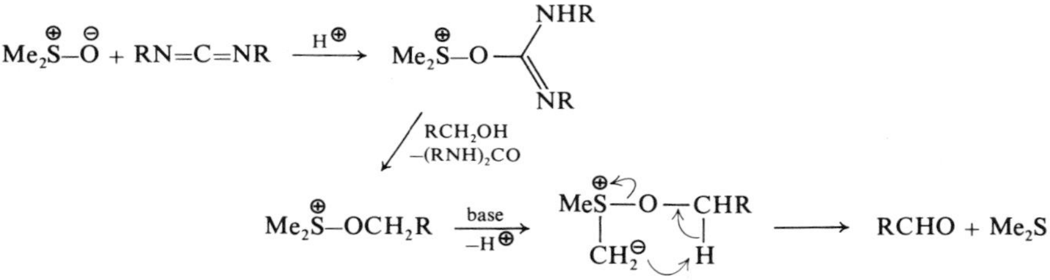

Barton *et al.* (134) achieved the alcohol oxidation by converting the substrates into the chloroformate esters which are capable of acylating DMSO. Subsequently, Corey and Kim (135, 136) reported a much less laborious method utilizing halosulfonium salts derived from dimethyl sulfide. The reaction of these highly electrophilic species with alcohols produces the same dimethylsulfoxonium intermediates.

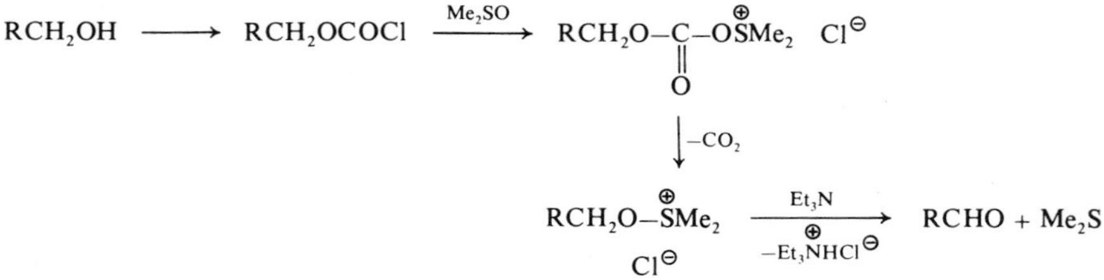

The effective reagents prepared from dimethyl sulfide and *N*-halosuccinimides (137) possess the general structure $(CH_2CO)_2N-\overset{\oplus}{S}Me_2\ X^{\ominus}$, which may be secondary products arising from a biphilic reaction. Tuleen and Buchanan

(138) interpreted the demethylimidation of aryl methyl sulfide with N-bromo-succinimide by such a pathway.

Sulfoxonium salts react with hard bases at sulfur and with soft nucleophiles at the oxygen end. Although the acid-catalyzed oxidation of isonitriles to isocyanates (139) has been formulated as involving an α addition of $H^{\oplus}$ and DMSO at the carbon atom of the isonitrile, the alternative mechanism shown below appears more viable in the HSAB context.

$$RN^{\oplus}{\equiv}C^{\ominus} + Me_2S^{\oplus}{-}OH \quad \underset{X^{\ominus}}{} \quad \xrightarrow{-Me_2S} \quad RN^{\oplus}{\equiv}C{-}OH \quad \underset{X^{\ominus}}{} \quad \xrightarrow{-HX} \quad RN{=}C{=}O$$

The alkylative hydration of nitriles is a quite different matter. The role of DMSO is that of a hard base for neutralizing a very hard nitrilium carbon. The synthesis of N-tritylamides (140) is representative.

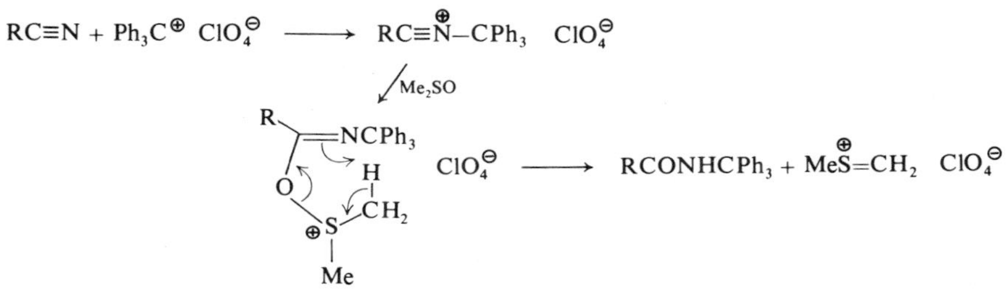

$$RC{\equiv}N + Ph_3C^{\oplus}\ ClO_4^{\ominus} \longrightarrow RC{\equiv}\overset{\oplus}{N}{-}CPh_3\ \ ClO_4^{\ominus}$$

$$\longrightarrow RCONHCPh_3 + Me\overset{\oplus}{S}{=}CH_2\ \ ClO_4^{\ominus}$$

Sulfoxides may be α-chlorinated by iodobenzene dichloride. Steps involving S–I bond formation, deprotonation, and internal chlorine transfer are postulated to account for the overall process (141). The iodine atom in the reagent plays the dual roles of an anchoring group as well as of an activator toward ylide generation.

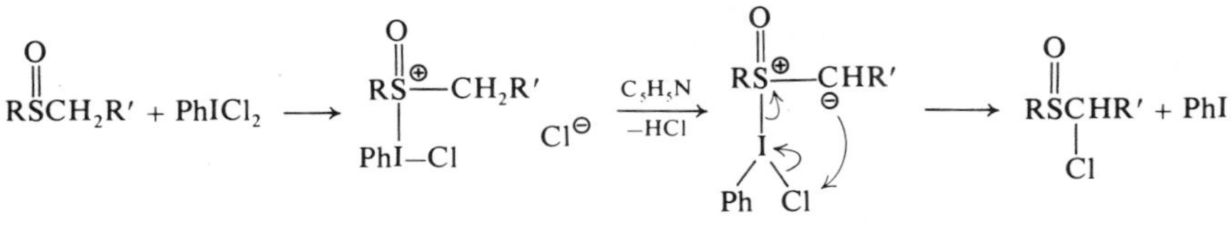

A sulfoxide deoxygenation method consists of treatment with dichloroborane (142). The substitution of two hydrogen atoms of borane by chlorine appears to impart the boron with a hardness compatible with the sulfoxide oxygen.

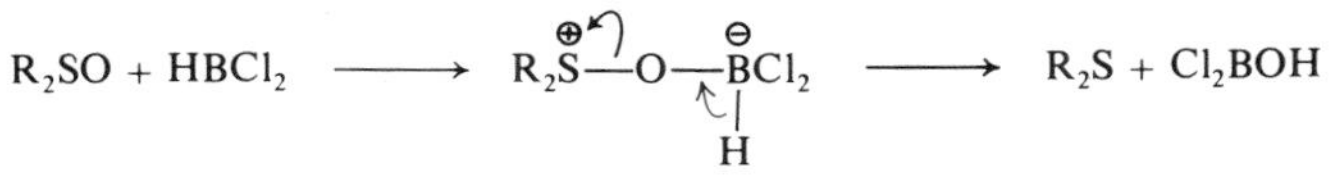

$$R_2SO + HBCl_2 \longrightarrow R_2\overset{\oplus}{S}{-}O{-}\overset{\ominus}{B}Cl_2 \longrightarrow R_2S + Cl_2BOH$$

Another procedure for the deoxygenation calls for the use of phenyl phenylenedioxyphosphine (143). However, the mechanism presented by the authors does not explain the acceleration effect of iodine and the fact that aliphatic sulfoxides react faster than their aromatic counterparts. The alternative pathways below agree better with experimental observations.

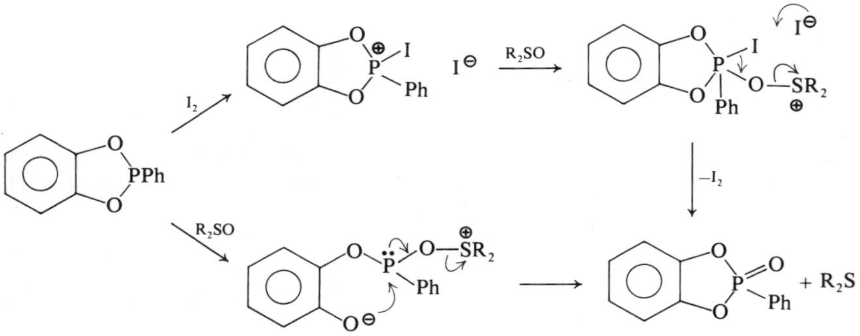

The acid-catalyzed reactions of sulfoxides with halide ions in protic solvents involve protonated species (144). The iodide ion effects simple reduction, presumably via bonding with the soft $OH^{\oplus}$. The harder $Br^{\ominus}$ and $Cl^{\ominus}$ promote racemization of optically active sulfoxides by the formation of sulfurane intermediates.

Alkoxysulfonium salts are cleaved by bromide and iodide ions (145) to sulfoxides with retained configurations indicating a $S_N2(C)$ displacement at alkoxy carbon by the soft nucleophiles. The use of equimolar chloride ion leads to a mixture of sulfides and partially racemized sulfoxides. The intervention of sulfurane intermediates has been postulated (146, 147) for the reaction.

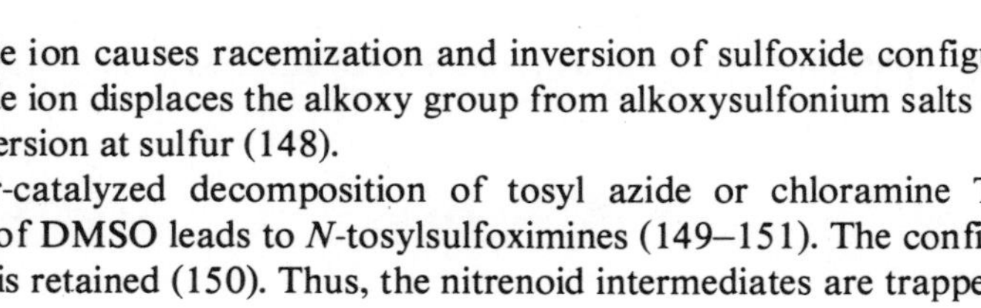

Fluoride ion causes racemization and inversion of sulfoxide configurations. Hydroxide ion displaces the alkoxy group from alkoxysulfonium salts resulting in net inversion at sulfur (148).

Copper-catalyzed decomposition of tosyl azide or chloramine T in the presence of DMSO leads to N-tosylsulfoximines (149–151). The configuration of sulfur is retained (150). Thus, the nitrenoid intermediates are trapped by the softer sulfur of the sulfoxides. A similar interception of nitrenes has also been reported (152, 153).

$$R_2SO + TsNXY \xrightarrow[\Delta]{Cu} R_2S\overset{O}{\underset{NTs}{\diagup}} + XY$$

Heating arenesulfinyl azides in DMSO affords arenesulfonyl sulfilimines (154). The results are consistent with nitrene trapping followed by rearrangement.

Carbenes are also intercepted by sulfoxides. For example, diazo compounds give sulfinyl ylides (155) during their decomposition by cuprous cyanide in the presence of a sulfoxide.

$$R_2SO + R'_2C=N_2 \xrightarrow[-N_2]{CuCN} R_2\overset{\oplus}{\underset{\parallel}{S}}-\overset{\ominus}{C}R'_2$$
$$O$$

Sulfinamides are chlorinated at the soft sulfur atom by 1-chlorobenzotriazole (156) to yield sulfonimidoyl chlorides.

9.5. REACTIONS OF HIGHER VALENT SULFUR DERIVATIVES

The sulfonyl group (sulfinate) is an ambident nucleophile which shows a discriminative mode of action toward different kinds of electrophiles. The methylation of sodium p-toluenesulfinate (157) affords predominantly a sulfinate ester when the leaving group is hard; it gives methyl p-tolylsulfone when reacting with soft reagents.

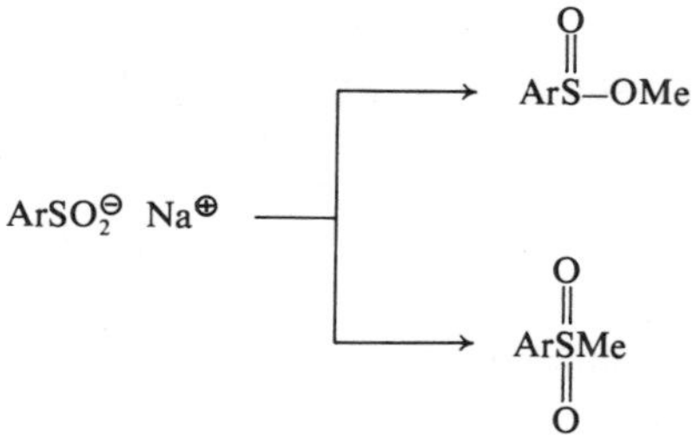

$$ArSO_2^{\ominus}\ Na^{\oplus}$$

$$\begin{array}{c} O \\ \parallel \\ ArS-OMe \end{array}$$

$$\begin{array}{c} O \\ \parallel \\ ArSMe \\ \parallel \\ O \end{array}$$

The alkylation of sulfinate anion by haloforms is possible only in the presence of aqueous alkali (158). Dihalocarbenes must be the true electrophiles. The exclusive formation of sulfones by this route provides support for the HSAB principle.

$$RSO_2^{\ominus} \; Na^{\oplus} + CHX_3 \xrightarrow{\;OH^{\ominus}\;} R-\underset{\underset{O}{\|}}{\overset{\overset{O}{\|}}{S}}-CHX_2$$

$$X = Cl, Br$$

p-Toluenesulfinate anion adds to α,β- and β,β-dichloroacrylonitriles in the Michael fashion using the sulfur as the attacking point (159). This is characteristic of a soft–soft interaction.

A sulfur equivalent of the Perkov reaction has been reported (160). Whether the sulfinate ion attacks on the halogen (biphilic) or directly on the carbonyl oxygen remains to be clarified.

$$ArC-C(COOR)_2 + PhSO_2^{\ominus} \; Na^{\oplus} \longrightarrow \quad + NaBr$$

During the reduction of aromatic diazonium salts (161) to arylhydrazines, a successive addition of the soft sulfite anion to the two nitrogen atoms is most likely involved. The nitrogen acceptors are soft.

$$ArN_2^{\oplus} \; Cl^{\ominus} \xrightarrow[-Cl^{\ominus}]{SO_3^{2\ominus}} ArN=N-SO_3^{\ominus} \xrightarrow[H^{\oplus}]{SO_3^{2\ominus}} \underset{\underset{SO_3^{\ominus}}{|}}{ArN-NHSO_3^{\ominus}} \xrightarrow{H^{\oplus}} ArNHNH_3^{\oplus}$$

It should be noted that hard bases prefer other pathways such as removing an *ortho* hydrogen resulting in benzyne (162, 163).

Benzenesulfonyl azide transfers its azido function to Grignard reagents (164). The soft reagents do not attack the hard sulfonyl group.

Kice and co-workers have investigated systematically the electrophilic behavior of sulfur atoms in the various oxidation states. As anticipated, it is found that the hardness increases in the order sulfenyl < sulfinyl < sulfonyl (165, 166). The sulfenyl S is a typical soft acceptor, whereas the sulfinyl group is medium soft and approximates the tetrahedral carbon. The sulfonyl S is as hard as the carbonyl of esters (167). The opposite characteristics of sulfinyl and sulfonyl S are qualitatively predicted, as the sulfinyl S has a lone-pair of electrons, whereas the sulfonyl counterpart does not, and there is a significantly higher positive charge residing on the sulfonyl sulfur than the sulfinyl S atom.

A dissentient view regarding the relative hardness of sulfonyl sulfur has been expressed by Rogne (168) who considers that its right place should be in between the carbonyl and the tetrahedral carbon atoms. In any event, Kice's generalization is useful for rationalizing the successful conversion of aryl chlorosulfates into the fluorosulfates (169) by the action of KF, and the tauto-merism (170) shown below. Since sulfinyl S is quite soft, the borderline chloride is a better bonding partner than the alkoxy oxygen. Note that in the carbon analog only the phthalide structure represents the stable entity.

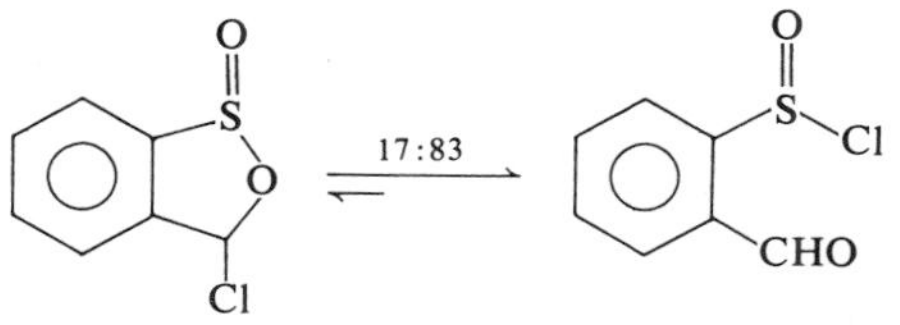

Chemical reduction of the sulfonyl group is not easy because most reducing agents are soft. The low yield conversion of sulfones to sulfides by LiAlH$_4$ (171) is improved by using i-Bu$_2$AlH (172). The latter reagent is harder and its uncharged Al atom permits interaction with the hard oxygen atoms of the sulfonyl group. Sodium diethylaluminum hydride is ineffective as it is even softer than LiAlH$_4$.

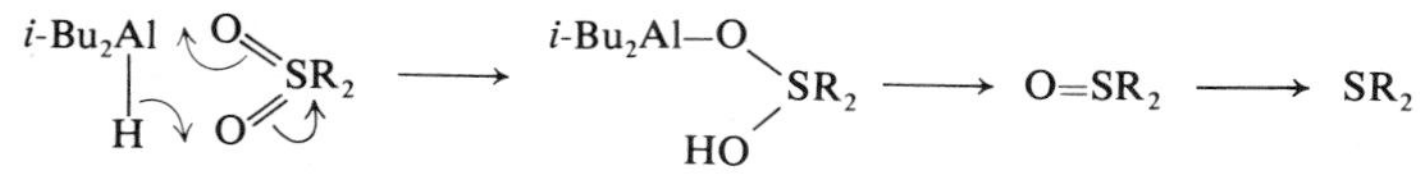

The release of Cl$^\ominus$ and PhO$^\ominus$ from phenyl chlorosulfate is effected by treatment with methoxide ion (173). The observed temperature dependence indicates that the O–S bond fission has a much higher enthalpy of activation than that leading to S–Cl cleavage, as symbiosis would suggest.

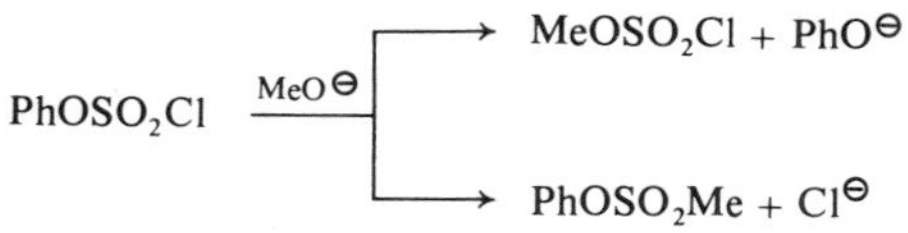

The cheletropic reaction* between conjugated dienes and sulfur dioxide yields cyclic sulfones. However, the isolated adducts of selenium dioxide and dienes possess the selenalactone structures (174). The only apparent explanation for the different stabilities of the cheletropic products (175) is that selenium is softer and tends to avoid a higher oxidation state.

* Cheletropic reactions are one kind of biphilic processes.

Sharpless and Lauer (176) have presented evidence in favor of allylseleninic acids as intermediates during oxidation of alkenes by selenium dioxide. Sulfur dioxide does not undergo this type of reaction with simple alkenes.

The structure of the oxidation product of 2,5-dihydro[c]benzoselenophene has been revised to that of an aldehydic diselenide (177) and not a selenaphthalide as previously claimed. The latter structure, which contains a chemical bond connecting a rather hard carbonyl carbon to a soft selenium atom, is evidently destabilized.

REFERENCES

1. M. L. Wolfrom, *J. Am. Chem. Soc.* **51**, 2188 (1929).
2. D. Seebach, *Synthesis* p. 17 (1969).
3. E. Vedejs and P. L. Fuchs, *J. Org. Chem.* **36**, 366 (1971).
4. T.-L. Ho and C. M. Wong, *Can. J. Chem.* **50**, 3740 (1972).
5. D. Gravel, C. Vaziri, and S. Rahal, *Chem. Commun.* p. 1323 (1972).
6. T. Mukaiyama, K. Maekawa, and K. Narasaka, *Chem. Lett.* p. 273 (1972).
7. K. Narasaka, T. Sakashita, and T. Mukaiyama, *Bull. Chem. Soc. Jpn.* **45**, 3724 (1972).
8. T. Oishi, K. Kamamoto, and Y. Ban, *Tetrahedron Lett.* p. 1085 (1972).
9. M. Fetizon and M. Jurion, *Chem. Commun.* p. 382 (1972).
10. T.-L. Ho and C. M. Wong, *Synthesis* p. 561 (1972).
11. H.-L. W. Chang, *Tetrahedron Lett.* p. 1989 (1972).
12. E. J. Corey and B. W. Erickson, *J. Org. Chem.* **36**, 3553 (1971).
13. Y. Tamura, K. Sumoto, S. Fujii, H. Satoh, and M. Ikeda, *Synthesis* p. 312 (1973).
14. W. F. J. Huurdeman, H. Wynberg, and D. W. Emerson, *Tetrahedron Lett.* p. 3449 (1971).
15. F. Guinot and G. Lamaty, *Tetrahedon Lett.* p. 2569 (1972); K. Pihlaja, *J. Am. Chem. Soc.* **94**, 3330 (1972).
16. F. Guinot, G. Lamaty, and H. Munsch, *Bull. Soc. Chim. Fr.* p. 541 (1971).
17. J. J. Delpuech and D. Nicole, *J. Chem. Soc., Perkin Trans. 2* p. 1025 (1974).

18. E. J. Corey and K. C. Nicolaou, *J. Am. Chem. Soc.* **96**, 5614 (1974).
19. S. Masamune, S. Kamata, and W. Schilling, *J. Am. Chem. Soc.* **97**, 3515 (1975).
20. S. Masamune, H. Yamamoto, S. Kamata, and A. Fukuzawa, *J. Am. Chem. Soc.* **97**, 3515 (1975).
21. R. Schwyzer and C. Hurlimann, *Helv. Chim. Acta* **37**, 155 (1954).
22. G. Wagner and P. Nuhn, *Arch. Pharm. (Weinheim, Ger.)* **298**, 686 (1965).
23. G. Wagner and P. Nuhn, *Arch. Pharm. (Weinheim, Ger.)* **298**, 692 (1965).
24. D. R. Hogg and A. Robertson, *Tetrahedron Lett.* p. 3783 (1974).
25. O. Foss, *Acta Chem. Scand.* **1**, 307 (1947).
26. J. Michalski, *Rocz. Chem.* **29**, 960 (1955).
27. C. K. Tseng and J. H.-H. Chan, *Tetrahedron Lett.* p. 699 (1971).
28. N. Watanabe, M. Okano, and S. Uemura, *Bull. Chem. Soc. Jpn.* **47**, 2745 (1974).
29. U. Tonellato, *Boll. Sci. Fac. Chim. Ind. Bologna* **27**, 249 (1969).
30. U. Tonellato and G. Levorato, *Boll. Sci. Fac. Chim. Ind. Bologna* **27**, 261 (1969).
31. K. R. H. Wooldridge and R. Slack, *J. Chem. Soc.* p. 1863 (1962).
32. D. Hoppe, *Angew. Chem., Int. Ed. Engl.* **12**, 656 (1973).
33. D. Hoppe, *Angew. Chem., Int. Ed. Engl.* **12**, 923 (1973).
34. D. Hoppe, *Angew. Chem., Int. Ed. Engl.* **12**, 658 (1973).
35. A. F. Halasa and G. E. P. Smith, *J. Org. Chem.* **36**, 636 (1971).
36. G. Barnikow and G. Strickmann, *Chem. Ber.* **100**, 1428 (1967).
37. E. J. Corey and R. H. K. Chen, *Tetrahedron Lett.* p. 3817 (1973).
38. H. Ishihara and S. Kato, *Tetrahedron Lett.* p. 3751 (1972).
39. M. Mikolajczyk, *Chem. Ind. (London)* p. 2059 (1966).
40. J. Michalski and Z. Tulimowski, *Bull. Acad. Pol. Sci., Ser. Sci. Chim.* **14**, 217 (1966).
41. W. J. Stec, T. Sudol, and B. Uznanski, *Chem. Commun.* p. 467 (1975).
42. A. Anciaux, A. Eman, W. Dumont, D. VanEnde, and A. Krief, *Tetrahedron Lett.* p. 1613 (1975).
43. A. Anciaux, A. Eman, W. Dumont, and A. Krief, *Tetrahedron Lett.* p. 1617 (1975).
44. D. Seebach and E. J. Corey, *J. Org. Chem.* **40**, 231 (1975).
45. K. Kondo, N. Sonoda, and S. Tsutsumi, *Tetrahedron Lett.* p. 4885 (1971).
46. N. Sonoda, T. Yasuhara, K. Kondo, T. Ikeda, and S. Tsutsumi, *J. Am. Chem. Soc.* **93**, 6344 (1971).
47. K. Kondo, N. Sonoda, and S. Tsutsumi, *Chem. Commun.* p. 307 (1972); P. Koch and E. Perrotti, *Tetrahedron Lett.* p. 2899 (1974); N. Sonoda, G. Yamamoto, K. Natsukawa, K. Kondo, and S. Murai, *ibid.* p. 1969 (1975).
48. M. Oki, W. Funakoshi, and A. Nakamura, *Bull. Chem. Soc. Jpn.* **44**, 828 (1971).
49. T. Mukaiyama, K. Hagio, H. Takei, and K. Saigo, *Bull. Chem. Soc. Jpn.* **44**, 161 (1971).
50. T. Mukaiyama, T. Adachi, and T. Kumamoto, *Bull. Chem. Soc. Jpn.* **44**, 3155 (1971).
51. B. M. Trost and H. C. Arndt, *J. Org. Chem.* **38**, 3140 (1973).
52. Z. Yoshida, H. Ogoshi, and T. Tokumitsu, *Tetrahedron* **26**, 2987 (1970).
53. M. Roth, P. Dubs, E. Götschi, and A. Eschenmoser, *Helv. Chim. Acta* **54**, 710 (1971).
54. M. Oki and K. Kobayashi, *Bull. Chem. Soc. Jpn.* **43**, 1223 (1970).
55. S. Kawamura, T. Horii, and J. Tsurugi, *J. Org. Chem.* **36**, 3677 (1971).
56. D. L. J. Clive and C. V. Denyer, *Chem. Commun.* p. 773 (1972).
57. G. E. Wilson and R. Albert, *J. Org. Chem.* **38**, 2160 (1973).
58. J. M. McIntosh and H. B. Goodbrand, *Tetrahedron Lett.* p. 3157 (1973).
59. L. Horner and E. Jürgens, *Justus Liebigs Ann. Chem.* **602**, 135 (1957).
60. J. J. Pappas, W. P. Keaveney, E. Gancher, and M. Berger, *Tetrahedron Lett.* p. 4273 (1966).
61. S. Oae, A. Nakanishi, and N. Tsujimoto, *Tetrahedron* **28**, 2981 (1972).

62. B. S. Campbell, D. B. Denney, D. Z. Denney, and L.-s. Shih, *J. Am. Chem. Soc.* **97**, 3850 (1975).
63. W. E. Parham and S. H. Groen, *J. Org. Chem.* **29**, 2214 (1964).
64. W. Grimme, J. Reisdorff, W. Jünemann, and E. Vogel, *J. Am. Chem. Soc.* **92**, 6335 (1970).
65. R. A. Moss, *J. Am. Chem. Soc.* **94**, 6004 (1972).
66. R. A. Moss and C. B. Mallon, *J. Org. Chem.* **40**, 1368 (1975).
67. W. E. Parham and S. H. Groen, *J. Org. Chem.* **29**, 2214 (1964).
68. K. Kondo and I. Ojima, *Chem. Commun.* p. 62 (1972).
69. W. Ando, H. Fujii, T. Takeuchi, H. Higuchi, Y. Saiki, and T. Migita, *Tetrahedron Lett.* p. 2117 (1973).
70. Y. Hata, M. Watanabe, S. Inoue, and S. Oae, *J. Am. Chem. Soc.* **97**, 2553 (1975).
71. H. Nozaki, H. Takaya, and R. Noyori, *Tetrahedron* **22**, 3393 (1966).
72. S. Searles, Jr. and R. W. Wann, *Tetrahedron Lett.* p. 2899 (1965).
73. L. Field and C. H. Banks, *J. Org. Chem.* **40**, 2774 (1975).
74. U. Schmidt and C. Osterroht, *Angew. Chem., Int. Ed. Engl.* **4**, 437 (1965).
75. O. J. Sherer and G. Wolmershauser, *Angew. Chem.* **87**, 485 (1975).
76. R. B. Woodward, *Science* **153**, 487 (1966).
77. R. B. Woodward, K. Heusler, J. Gosteli, P. Naegeli, W. Oppolzer, R. Ramage, S. Ranganathan, and H. Vorbrüggen, *J. Am. Chem. Soc.* **88**, 852 (1966).
78. O. Mitsunobu, K. Kato, and M. Tomari, *Tetrahedron* **26**, 5731 (1970).
79. Y. Poirier and N. Lozac'h, *Bull. Soc. Chim. Fr.* p. 2090 (1967).
80. D. Gonbeau and G. Pfister-Guillouzo, *Tetrahedron* **31**, 459 (1975).
80a. R. E. Davis, H. Nakshbendi, and A. Ohno, *J. Org. Chem.* **31**, 2702 (1966).
81. C. A. Grob, B. Schmitz, A. Sutter, and A. H. Weber, *Tetrahedron Lett.* p. 3551 (1975).
82. G. K. Helmkamp and D. J. Pettitt, *J. Org. Chem.* **27**, 2942 (1962).
83. G. K. Helmkamp and D. J. Pettitt, *J. Org. Chem.* **25**, 1754 (1960).
84. D. B. Denney and M. J. Boskin, *J. Am. Chem. Soc.* **82**, 4736 (1960).
85. N. P. Neureiter and F. G. Bordwell, *J. Am. Chem. Soc.* **81**, 578 (1959).
86. Y. Kobayashi, I. Kumadaki, A. Ohsawa, and Y. Sekine, *Tetrahedron Lett.* p. 1639 (1975).
87. F. G. Bordwell, H. M. Anderson, and B. M. Pitt, *J. Am. Chem. Soc.* **76**, 1082 (1954).
88. D. A. Lightner and C. Djerassi, *Chem. Ind.* (*London*) p. 1236 (1962); K. Takeda and T. Komeno, *ibid.* p. 1793 (1962).
89. D. Seebach, *Angew. Chem., Int. Ed. Engl.* **6**, 442 (1967); D. Seebach, *ibid.* **6**, 443 (1967).
90. T. S. Woods and D. L. Klayman, *J. Org. Chem.* **39**, 3716 (1974).
91. R. A. Ellison, W. D. Woessner, and C. C. Williams, *J. Org. Chem.* **37**, 2757 (1972).
92. O. Foss, *Acta Chem. Scand.* **1**, 307 (1947).
93. H. B. Footner and S. Smiles, *J. Chem. Soc.* **127**, 2887 (1925).
94. R. D. Ritter and J. H. Krueger, *J. Am. Chem. Soc.* **92**, 2316 (1970).
95. S. J. Brois, J. F. Pilot, and H. W. Barnum, *J. Am. Chem. Soc.* **92**, 7629 (1970).
96. D. N. Harpp and A. Granata, *Tetrahedron Lett.* p. 3001 (1976).
97. T. Kumamoto, S. Kobayashi, and T. Mukaiyama, *Bull. Chem. Soc. Jpn.* **45**, 866 (1972).
98. K. S. Boustany and A. B. Sullivan, *Tetrahedron Lett.* p. 3547 (1970).
99. D. N. Harpp, D. K. Ash, T. G. Back, J. G. Gleason, B. A. Orwig, W. F. VanHorn, and J. P. Snyder, *Tetrahedron Lett.* p. 3551 (1970).
100. D. N. Harpp and T. G. Back, *J. Org. Chem.* **36**, 3828 (1971); *Tetrahedron Lett.* p. 4953 (1971).
101. M. Furukawa, T. Suda, A. Tsukamoto, and S. Hayashi, *Synthesis* p. 165 (1975).
102a. T. Mukaiyama and T. Taguchi, *Tetrahedron Lett.* p. 3411 (1970).
102b. T. Mukaiyama and K. Takahashi, *Tetrahedron Lett.* p. 5907 (1968).
103. D. N. Harpp and B. A. Orwig, *Tetrahedron Lett.* p. 2691 (1970).

104. F. Becke and H. Hagen, *Justus Liebigs Ann. Chem.* **729**, 146 (1969).
105. D. E. L. Carrington, K. Clarke, and R. M. Scrowston, *J. Chem. Soc. C*, pp. 3262 and 3903 (1971).
106. C. E. Diebert, *J. Org. Chem.* **35**, 1501 (1970).
107. Yu. A. Sharanin, *Kratk. Tezisy-Vses. Soveshch. Probl. Mekh. Geteroliticheskikh Reakts.* p. 153 (1974).
108. A. Schönberg, B. Konig, and E. Singer, *Chem. Ber.* **100**, 767 (1967).
109. W. J. Middleton and W. H. Sharkey, *J. Org. Chem.* **30**, 1384 (1965).
110. E. J. Corey and G. Märkl, *Tetrahedron Lett.* p. 3201 (1967).
111. E. J. Corey and R. A. E. Winter, *J. Am. Chem. Soc.* **85**, 2677 (1963).
112. Z. Yoshida, T. Kawase, and S. Yoneda, *Tetrahedron Lett.* p. 235 (1975).
113. A. Schönberg, E. Singer, E. Frese, and K. Praefcke, *Chem. Ber.* **98**, 3311 (1965).
114. P. Beak and J. W. Worley, *J. Am. Chem. Soc.* **92**, 4142 (1970).
115. P. Beak and J. W. Worley, *J. Am. Chem. Soc.* **94**, 597 (1972).
116. M. Dagonneau, J.-F. Hemidy, D. Cornet, and J. Vialle, *Tetrahedron Lett.* p. 3003 (1972).
117. R. F. Hudson, *Angew. Chem., Int. Ed. Engl.* **12**, 36 (1973).
118. M. Dagonneau and J. Vialle, *Tetrahedron* **30**, 3119 (1974).
119. L. Leger and M. Saquet, *Bull. Soc. Chim. Fr.* p. 657 (1975).
120. K. Diemert and W. Kuchen, *Angew. Chem., Int. Ed. Engl.* **10**, 508 (1971).
121. S. G. Smith and S. Winstein, *Tetrahedron* **3**, 317 (1958).
122. R. Kuhn and H. Trischmann, *Justus Liebigs Ann. Chem.* **611**, 117 (1958).
123. A. P. Johnson and A. Pelter, *J. Chem. Soc.* p. 520 (1964).
124. W. W. Epstein and J. Ollinger, *Chem. Commun.* p. 1338 (1970).
125. D. M. Lemal and A. J. Fry, *J. Org. Chem.* **29**, 1673 (1964).
126. B. Ganem and R. K. Boeckman, Jr., *Tetrahedron Lett.* p. 917 (1974).
127. N. Kornblum, W. J. Jones, and G. J. Anderson, *J. Am. Chem. Soc.* **81**, 4113 (1959).
128. N. Kornblum, J. W. Powers, G. J. Anderson, W. J. Jones, H. O. Larson, O. Levand, and W. M. Weaver, *J. Am. Chem. Soc.* **79**, 6562 (1957).
129. I. M. Hunsberger and J. M. Tien, *Chem. Ind. (London)* p. 88 (1959); D. N. Kevill and V. V. Likhite, *Chem. Commun.* p. 247 (1967).
130. T. Cohen and T. Tsuji, *J. Org. Chem.* **26**, 1681 (1961).
131. K. E. Pfitzner and J. G. Moffatt, *J. Am. Chem. Soc.* **87**, 5661 and 5670 (1965).
132. A. H. Fenselau and J. G. Moffatt, *J. Am. Chem. Soc.* **88**, 1762 (1966).
133. W. W. Epstein and F. W. Sweat, *Chem. Rev.* **67**, 247 (1967).
134. D. H. R. Barton, B. J. Garner, and R. H. Wightman, *J. Chem. Soc.* p. 1855 (1964).
135. E. J. Corey and C. U. Kim, *J. Am. Chem. Soc.* **94**, 7586 (1972).
136. E. J. Corey and C. U. Kim, *Tetrahedron Lett.* p. 287 (1974).
137. E. Vilsmaier and W. Sprügel, *Justus Liebigs Ann. Chem.* **747**, 151 (1971).
138. D. L. Tuleen and D. N. Buchanan, *J. Org. Chem.* **32**, 495 (1967).
139. D. Martin and A. Weise, *Angew. Chem., Int. Ed. Engl.* **6**, 168 (1967).
140. D. Martin and A. Weise, *Justus Liebigs Ann. Chem.* **702**, 86 (1967).
141. M. Cinquini, S. Colonna, and F. Montanari, *Chem. Commun.* p. 607 (1969).
142. H. C. Brown and N. Ravindran, *Synthesis* p. 42 (1973).
143. M. Dreux, Y. Leroux, and P. Savignac, *Synthesis* p. 506 (1974).
144. D. Landini, G. Modena, F. Montanari, and G. Scorrano, *J. Am. Chem. Soc.* **92**, 7168 (1970).
145. R. Annunziata, M. Cinquini, and S. Colonna, *J. Chem. Soc., Perkin Trans. 1*, p. 404 (1975).
146. R. H. Rymbrandt, *Tetrahedron Lett.* p. 3553 (1971).
147. K. Mislow, *Rec. Chem. Prog.* **28**, 217 (1967).

148. C. R. Johnson and D. McCants, *J. Am. Chem. Soc.* **87**, 5404 (1965).
149. H. Kwart and A. A. Kahn, *J. Am. Chem. Soc.* **89**, 1950 (1967).
150. D. J. Cram, J. Day, D. R. Rayner, D. M. von Schriltz, D. J. Duchamp, and D. C. Garwood, *J. Am. Chem. Soc.* **92**, 7369 (1970).
151. C. R. Johnson, R. A. Kirchhoff, R. J. Reischer, and G. F. Katekar, *J. Am. Chem. Soc.* **95**, 4287 (1973).
152. J. Sauer and K. K. Mayer, *Tetrahedron Lett.* p. 319 (1968).
153. Y. Yamada, T. Yamamoto, and M. Okawara, *Chem. Lett.* p. 361 (1975).
154. T. J. Maricich and V. L. Hoffman, *Tetrahedron Lett.* p. 729 (1971).
155. F. Dost and J. Gosselck, *Chem. Ber.* **105**, 948 (1972).
156. E. U. Jonsson, C. C. Bacon, and C. R. Johnson, *J. Am. Chem. Soc.* **93**, 5306 (1971).
157. J. S. Meek and J. S. Fowler, *J. Org. Chem.* **33**, 3422 (1968).
158. W. Middelbos, J. Strating, and B. Zwanenburg, *Tetrahedron Lett.* p. 351 (1971).
159. M. V. Kalnins and B. Miller, *Chem. Ind. (London)* p. 555 (1966).
160. I. Fleming and C. R. Owen, *Chem. Commun.* p. 1402 (1970).
161. R. Huisgen and R. Lux, *Chem. Ber.* **93**, 540 (1960).
162. C. Rüchardt and C. C. Tan, *Angew. Chem., Int. Ed. Engl.* **9**, 522 (1970).
163. J. I. G. Cadogan, J. R. Mitchell, and J. T. Sharp, *Chem. Commun.* p. 1 (1971).
164. S. Ito, *Bull. Chem. Soc. Jpn.* **39**, 635 (1966).
165. J. L. Kice and G. Guaraldi, *J. Am. Chem. Soc.* **90**, 4076 (1968).
166. J. L. Kice and G. B. Large, *J. Am. Chem. Soc.* **90**, 4069 (1968).
167. J. L. Kice, G. J. Kasperek, and D. Patterson, *J. Am. Chem. Soc.* **91**, 5516 (1969).
168. O. Rogne, *J. Chem. Soc. B* p. 1056 (1970).
169. M. Hedayatullah, A. Guy, and L. Denivelle, *C. R. Hebd Seamces Acad. Sci., Ser. C* **278**, 57 (1974).
170. J. F. King, A. Hawson, B. L. Huston, L. J. Danks, and J. Komery, *Can. J. Chem.* **49**, 943 (1971).
171. F. G. Bordwell and W. H. McKellin, *J. Am. Chem. Soc.* **73**, 2251 (1951).
172. J. N. Gardner, S. Kaiser, A. Krubiner, and H. Lucas, *Can. J. Chem.* **51**, 1419 (1973).
173. E. Buncel, L. I. Choong, and A. Raoult, *J. Chem. Soc., Perkin Trans. 2* p. 691 (1972).
174. W. L. Mock and J. W. McCausland, *Tetrahedron Lett.* p. 391 (1968).
175. R. F. Heldeweg and H. Hogeveen, *J. Am. Chem. Soc.* **98**, 2341 (1976).
176. K. B. Sharpless and R. F. Lauer, *J. Am. Chem. Soc.* **94**, 7154 (1972).
177. B. E. Norcross and R. L. Martin, *Abstr. 169th Meet., Am. Chem. Soc., Philadelphia, 1975,* ORGN 93 (1975).

10

Organoboron Chemistry

10.1. STABILITY OF ORGANOBORANES. EXCHANGE REACTIONS

According to Ahrland *et al.* (1), boron is of borderline softness. Borane derivatives are mildly soft (class *b*) if all the ligands are from Groups V and VI. However, boron trifluoride is a hard acid.

10.1.1. Boron Trihalides

Boron trifluoride complexes easily with ethers. The complexes are stabilized by symbiosis of F and O ligands around the boron. Dialkyl ethers are ruptured by BBr_3 to furnish alkyl bromides (2, 3).* This is a consequence of the mutual weakening of B–Br and O–R bonds (both are hard–soft pairs) on coordination. The splitting of the bromide ion from the complexes and its return attack on the alkyl group of the oxonium intermediates are favored on HSAB grounds.

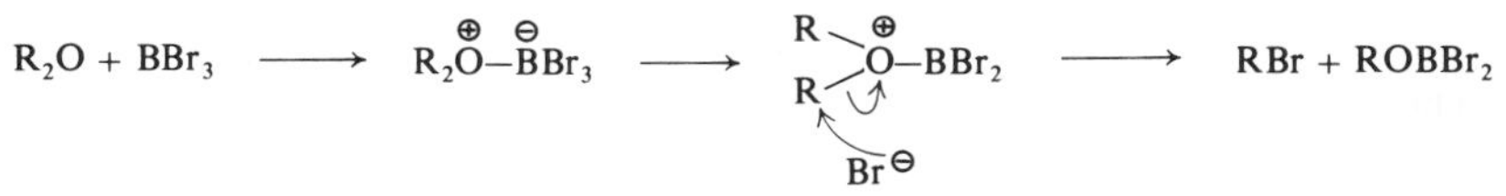

$$R_2O + BBr_3 \longrightarrow R_2\overset{\oplus}{O}-\overset{\ominus}{B}Br_3 \longrightarrow \overset{R}{\underset{R}{\diagdown}}\overset{\oplus}{O}-BBr_2 \longrightarrow RBr + ROBBr_2$$
$$\underset{Br^{\ominus}}{}$$

10.1.2. Thioboranes

Owing to its borderline softness, boron atom may interact equally well with both hard and soft bases. Thus, the solvolysis of thioboranes (5, 6) is easily ac-

---

* Different mechanisms operate in this reaction and in the cleavage of *sec-* and *tert-*alkyl ethers by BF_3 (4). The decomposition of the latter complexes proceeds by the S_N1 route.

complished. A simple admixture of thioboranes with carbonyl compounds affords dithioacetals (7, 8).

Mixed disulfides are obtained when sulfenate esters are treated with trialkylthioboranes (9). This reaction is very efficient because the sulfenates contain an unstable soft–hard bond.

$$3ArS-OMe + (RS)_3B \longrightarrow 3ArS-SR + (MeO)_3B$$

It is of interest to note that ethers are also cleaved on exposure to monoalkylthioboranes (10).

$$R-O-R + R'S-BH_2 \longrightarrow ROBH_2 + R'SR$$

An attempt to prepare mixed borates results in disproportionation products (11). Symbiosis appears to be the driving force for the secondary transformation.

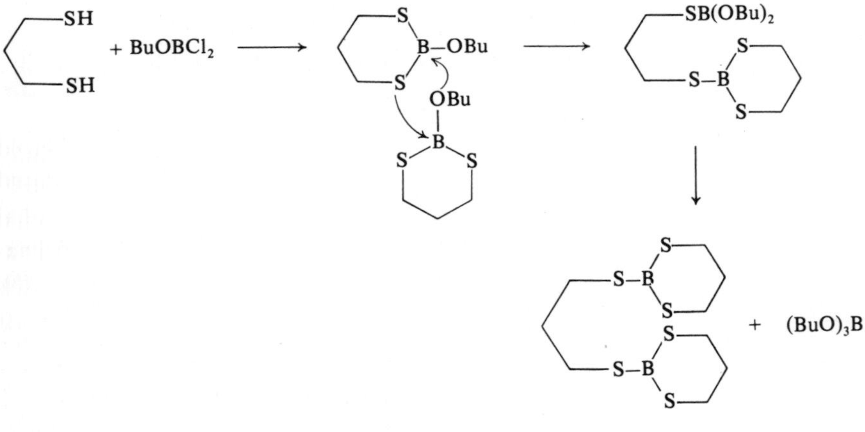

Thioboranes also undergo bromolysis (12) as shown below.

$$R_2B-SR' + Br_2 \longrightarrow R_2B-Br + R'S-Br$$

$$R_2B-SR' + R'S-Br \longrightarrow R_2B-Br + R'_2S_2$$

10.1.3. Oxygenative Degradation of Organoboranes

The synthetic utility of organoboranes depends on the feasibility of deboronative functionalization. Many methods are now known. The treatment of organoboranes with alkaline hydrogen peroxide leads to alcohols (13), this involves addition of $HO_2^\ominus$ to boron followed by migration of a soft alkyl group to the soft oxy center of the hydroperoxyboranide intermediates. Peracid (14), alkaline hypochlorite (15), and tertiary amine oxides (16, 17) may be employed as alternative reagents as they provide appropriate interacting loci for the organoboranes.

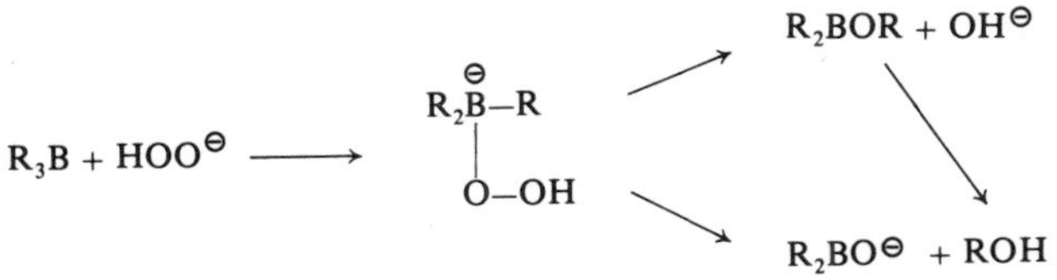

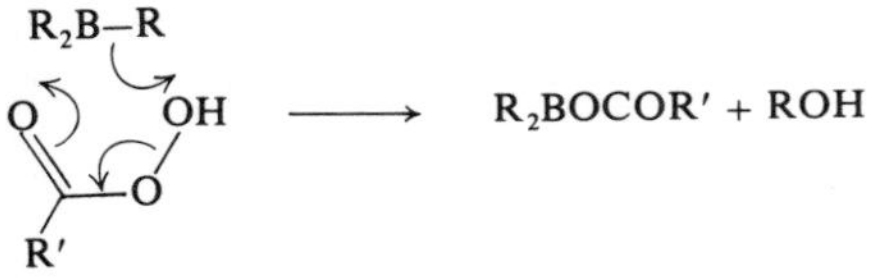

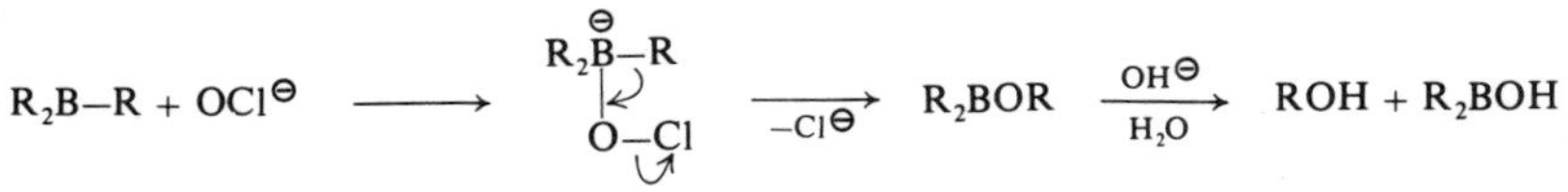

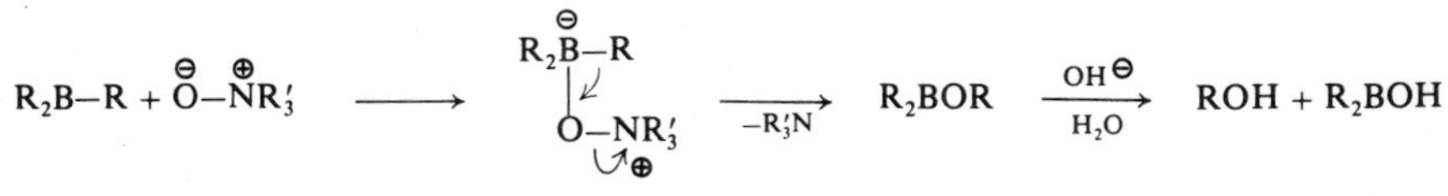

$$R_3B + HOO^\ominus \longrightarrow \begin{array}{c} R_2\overset{\ominus}{B}-R \\ | \\ O-OH \end{array} \nearrow \begin{array}{c} R_2BOR + OH^\ominus \\ \searrow \\ R_2BO^\ominus + ROH \end{array}$$

$$\begin{array}{c} R_2B-R \\ \diagdown \\ O \underset{R'}{\diagup} OH \\ \diagdown O \end{array} \longrightarrow R_2BOCOR' + ROH$$

$$R_2B-R + OCl^\ominus \longrightarrow \begin{array}{c} R_2\overset{\ominus}{B}-R \\ | \\ O-Cl \end{array} \xrightarrow{-Cl^\ominus} R_2BOR \xrightarrow[H_2O]{OH^\ominus} ROH + R_2BOH$$

$$R_2B-R + \overset{\ominus}{O}-\overset{\oplus}{N}R_3' \longrightarrow \begin{array}{c} R_2\overset{\ominus}{B}-R \\ | \\ O-\overset{\oplus}{N}R_3' \end{array} \xrightarrow{-R_3'N} R_2BOR \xrightarrow[H_2O]{OH^\ominus} ROH + R_2BOH$$

Degradation of benzeneboronic acid with bromine is greatly facilitated by added hard bases such as water (18). The findings suggest a push–pull mechanism in which the aromatic ring is activated by the hard donor which supplies electrons on bonding with the boron atom. Iododeboronation of naphthaleneboronic acids has also been studied (19).

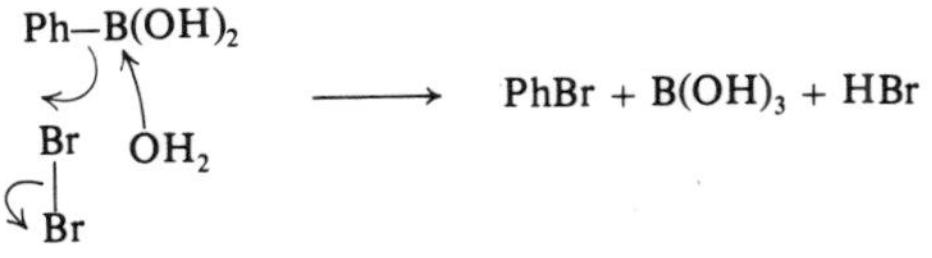

$$\begin{array}{c} Ph-B(OH)_2 \\ \text{Br} \quad \text{OH}_2 \\ | \\ \text{Br} \end{array} \longrightarrow PhBr + B(OH)_3 + HBr$$

10.1.4. Cleavage of C–B Bond by Other Nucleophiles

Chloramine and hydroxylamine-O-sulfonic acid react trialkylboranes to furnish alkylamines (20) via a 1,2-alkyl shift from boron to nitrogen. N-Chlorodialkylamines promote an exchange reaction (21) in the direction opposite to that effected by chloramine.

Homologous alcohols may be synthesized by treating trialkylboranes with dimethylsulfonium methylide (22) or dimethyloxosulfonium methylide (23), followed by oxidative workup.

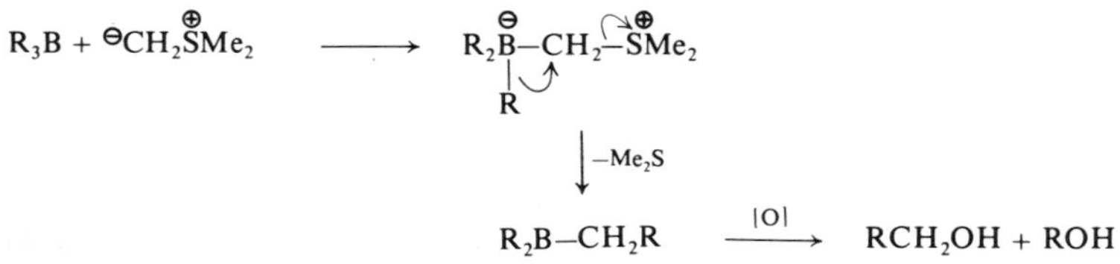

$$R_3B + {}^{\ominus}CH_2\overset{\oplus}{S}Me_2 \longrightarrow R_2\overset{\ominus}{B}-CH_2-\overset{\oplus}{S}Me_2$$
$$\qquad\qquad\qquad\qquad\qquad\qquad\quad |{}R$$

$$\downarrow {\scriptstyle -Me_2S}$$

$$R_2B-CH_2R \xrightarrow{\;|O|\;} RCH_2OH + ROH$$

Although carbenes are electron-deficient species, they react, as donors, with boranes. Secondary alcohols are produced via two successive alkyl migrations from the boron to carbenium centers of the intermediates when chlorocarbene (24) and methoxycarbene (25) are employed.

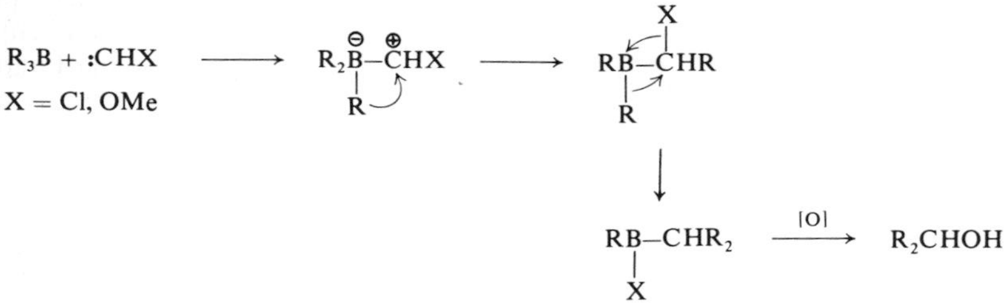

$$R_3B + :CHX \longrightarrow R_2\overset{\ominus}{B}-\overset{\oplus}{C}HX \longrightarrow R\overset{X}{\underset{R}{B}}-CHR$$
$$X = Cl, OMe$$

$$\downarrow$$

$$RB-CHR_2 \xrightarrow{\;|O|\;} R_2CHOH$$
$$|{}X$$

By having a carbanionoid center, diazo compounds react with boranes easily. For example, diazoketones yield β-oxoboranes which rearrange *in situ* to generate enol borates (26–28). The use of organoazides instead of diazo compounds leads to secondary amines (29, 30).

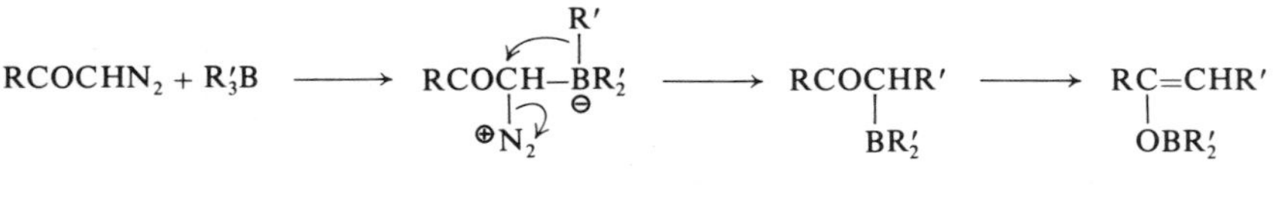

$$RCOCHN_2 + R_3'B \longrightarrow RCOCH-BR_2' \longrightarrow RCOCHR' \longrightarrow RC=CHR'$$
$$\qquad\qquad\qquad\qquad\quad |{}\overset{\oplus}{N_2}\;\;{}^{\ominus}\qquad\qquad\quad |{}BR_2'\qquad\qquad |{}OBR_2'$$

$$RN_3 + R_3'B \longrightarrow RR'N-BR_2' \xrightarrow{\;MeOH\;} RR'NH + R_2'BOMe$$

Indirect alkylation of ketones (31) and esters (32) via their α-bromo derivatives can be achieved by base-catalyzed reactions with trialkylboranes. The reaction is strictly analogous to the homologation of diazoketones.

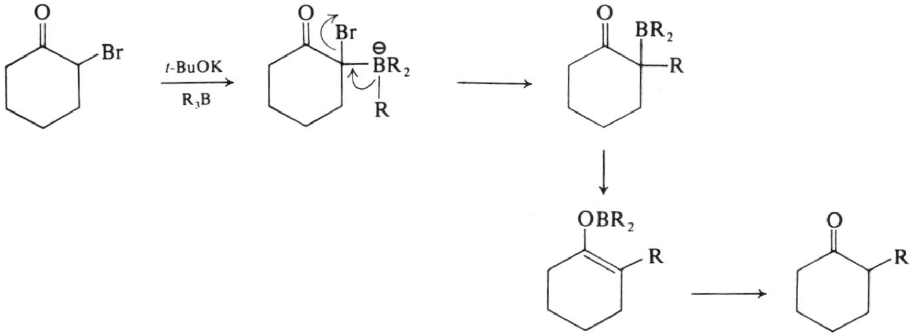

Triallylborane reacts with monosubstituted acetylenes by an ene-type process (33). This could be a reflection of the fact that boron prefers bonding to a slightly harder sp^2 carbon.

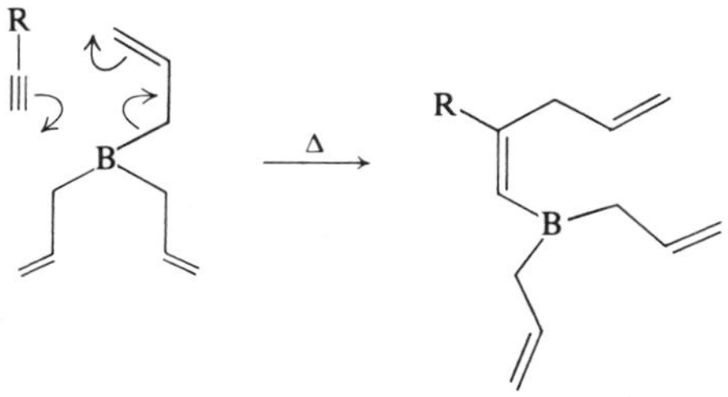

10.2. HYDROBORATION

Borane is a soft Lewis acid, therefore its complexation with soft olefin linkages is very favorable. Apparently the initial borane–alkene complexes collapse rapidly to a four-centered transition state leading to the organoboranes without the participation of external nucleophiles. This process is called hydroboration (34).

As far as the B–H moiety is concerned, the boron atom is the acceptor and the hydrogen (as $H^\ominus$) is a soft base. The unusual orientation of addition of the B–H elements across double bonds may be due to the thermodynamic stability of the transition state which has a certain degree of polar character.

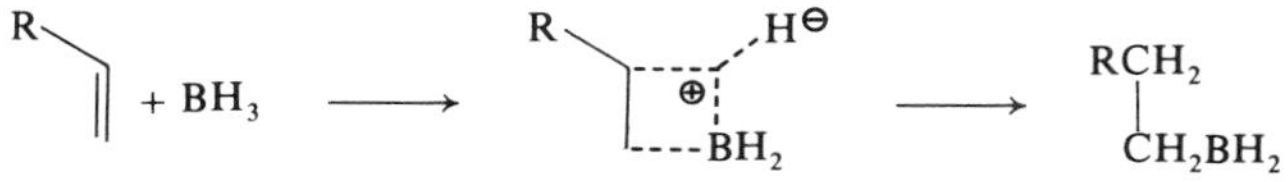

It has been found that the central atom of the allenic bond is the preferred boron-bonding site during hydroboration (35). This center is harder than the terminal sp^2 carbons.

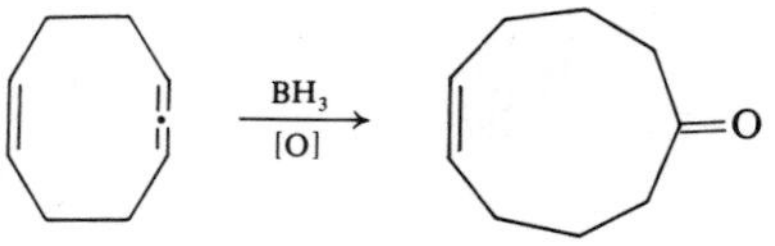

In the hydroboration of styrene derivatives, the orientation is influenced to some extent by the substituent on the benzene ring. The presence of an electron-withdrawing group stabilizes the transition state leading to boron attachment to the benzylic position. This carbon is harder than that of styrene. On the other hand, an electron-donating function tends to direct the C–B bonding toward the terminal carbon (36).

Vinylferrocene is hydroborated at the terminal position (37) to the extent of 98% as compared to 80% in the case of styrene. The greater softness of the ferrocene system versus the phenyl group is demonstrated by the regioselectivity of hydroboration.

Hydroboration of 1-haloalkenes (38, 39) and enol acetates places the boron atom predominantly at the α carbon. However, enol ethers (38) and enamines (40), regardless of the degree of substitution on the original double bond, give 1,2-disubstituted products.

$$RCH=CHX + R_2'BH \longrightarrow RCH_2\underset{\underset{X}{|}}{CHBR_2'}$$

$$R = Et, Ph$$
$$X = Cl, Br$$

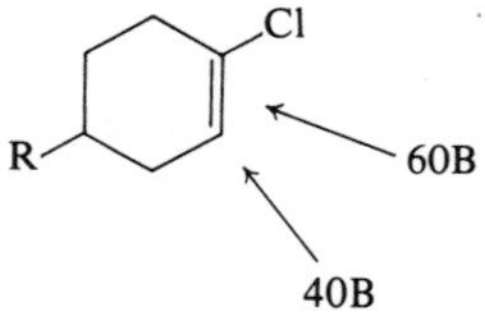

$$R = H, t\text{-Bu}$$

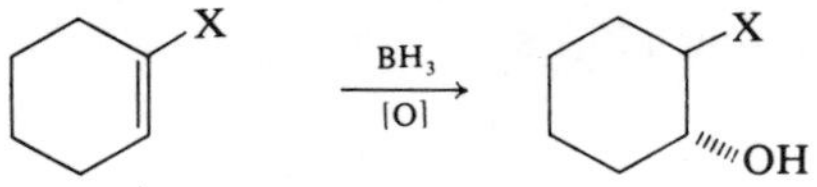

$$X = OEt, NC_4H_8$$

The halogen and the acetoxy group infuse hardness in the ipso carbon, rendering it a better partner for the boron. Although alkoxy [including intracyclic analogs (41) such as 2,3-dihydrofuran and dihydropyran] and amino substituents should exert a similar inductive effect, this is swamped by the resonance inherent in these systems. The contribution of the zwitterionic hybrid fixes the nucleophilic center at the β carbon.

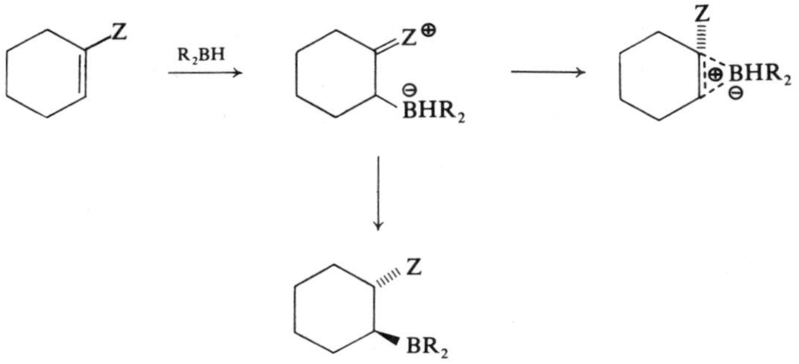

Chloroborane (ClBH$_2$) is a more regioselective hydroborating agent than borane (42). This property can be attributed to the heightened hardness of the boron atom.

The reaction of organoboranes with carbon–nitrogen multiple bonds always leads to aminoborane derivatives through a hard–hard interaction. Hydroboration of nitriles with optically active diisopinocampheylborane (43) constitutes a key step in an asymmetric synthesis of amino acids (44). Carbonyl compounds are deoxygenated via hydroboration of their tosylhydrazones (45).

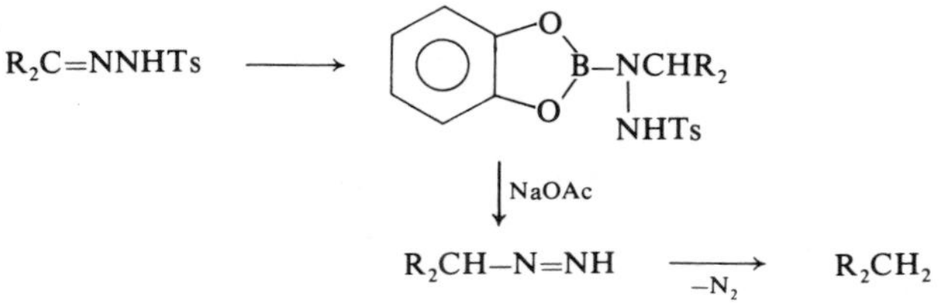

10.3. REACTIONS OF ARYLBORON COMPOUNDS

Arylboron compounds are susceptible to nucleophilic attack which often results in the cleavage of the C_{Ar}–B bond. On the other hand, the introduction of a boron substituent into the aromatic ring may be performed by the interaction of aryl iodides with boron triiodide (46, 47).

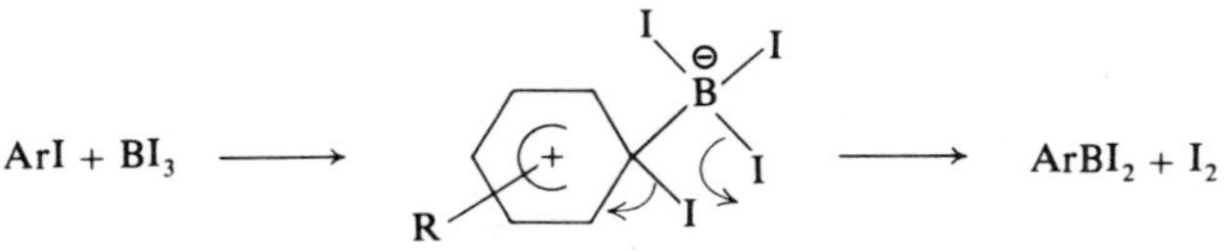

The formation of the σ intermediates is symbiotically favored as both iodine and BI₃ are soft. The decomposition of these intermediates is also a soft–soft process.

Ortho-disubstitution of aromatics via aryltrialkylboranides has been studied (48). Reaction of the boranides with alkyl fluorosulfates affords dihydroarene derivatives which rearomatize on treatment with alkaline hydrogen peroxide. The hydroperoxyboranides may decompose in either of two ways according to the HSAB concept.

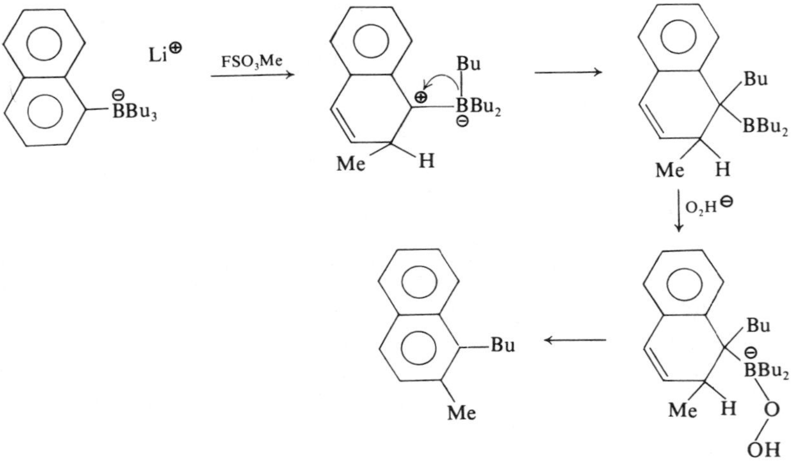

10.4. SOME ASPECTS OF ORGANOALUMINUM REACTIONS

Aluminum is a congener of boron, thus the two elements display similar chemical characteristics. One important difference is that boron is a metalloid, whereas aluminum is a *bona fide* metal. As a consequence, the boron center shows soft or hard characteristics depending on the ligands it carries; the aluminum center is hard most of the time.

Analogous to the reaction of carboxylic acids with thioboranes, thioesters are also synthesized from methyl esters by treatment with alkylthio-dimethylalanes (49, 50). Bifunctional reagents transform esters and lactones into ketene thioacetals and dithio orthoesters, respectively. In these reactions, the alkoxide moiety attaches to the hard Al atom. That the carbonyl C is left to combine with a soft sulfide represents a compromise as the hard–hard Al–O interaction far outweighs the loss in stability resultant from the change of
$-\underset{\underset{O}{\|}}{C}-O-$ to $-\underset{\underset{O}{\|}}{C}-S-$.

$$RCOOMe + R'SAlMe_2 \longrightarrow RCOSR' + MeOAlMe_2$$

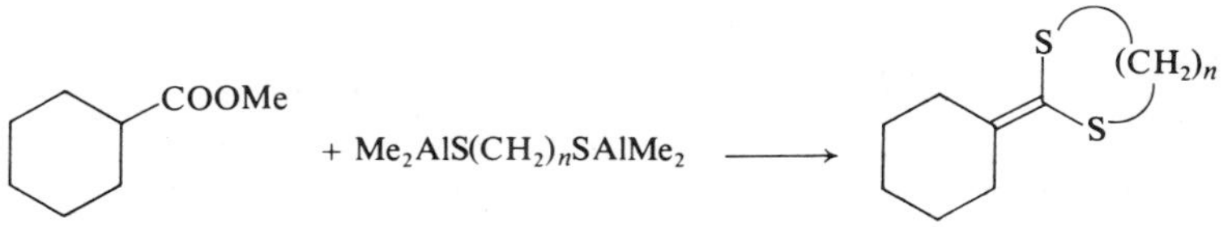

$$n = 2,3$$

Hydroalumination has received much less attention than hydroboration. The reduction of an isolated double bond in the LiAlH$_4$ reduction of ethyl 1-ethyl-cyclopropene-3-carboxylate (51) undoubtedly involves such a reaction.

Unsymmetrical epoxides are cleaved differently by lithium tetraalkyl-aluminate and trialkylaluminum (52). Alkyl transfer from the softer "ate" complexes takes place at the softer (less substituted) carbon atom, whereas the harder R$_3$Al reacts at the alternative position. It can also be shown that "ate" complexes effect opening of epoxides with inversion of configuration, whereas R$_3$Al gives alcohols in which the configurations are retained.

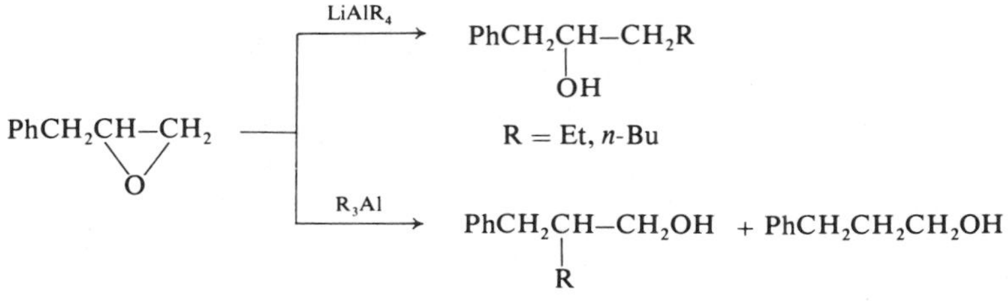

R = n-Bu	90%	10%
R = Et	80%	20%

The conversion of epoxides to allylic alcohols can be accomplished by dialkylamide anions, but the treatment with N-diethylalanyl-2,2,6,6-tetra-methylpiperidine (53) is more efficient. A cyclic syn-elimination pathway is indicated. A perfect match of hard interactions is provided by such combinations.

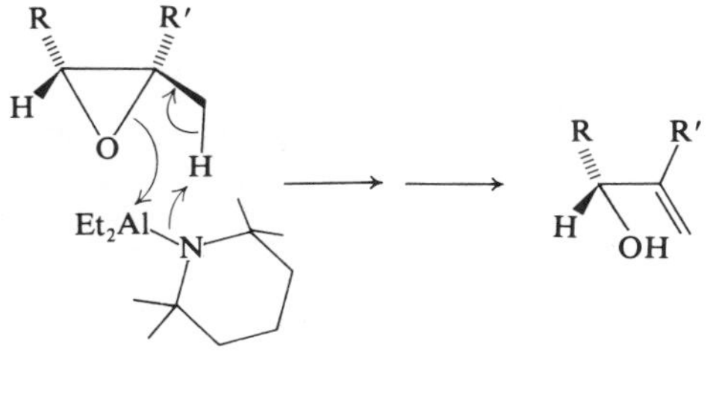

REFERENCES

1. S. Ahrland, J. Chatt, and N. R. Davies, *Q. Rev.*, **11**, 265 (1958).
2. F. L. Benton and T. E. Dillon, *J. Am. Chem. Soc.* **64**, 1128 (1942).
3. J. F. W. McOmie and M. L. Watts, *Chem. Ind.* (*London*) p. 1658 (1963).
4. E. F. Mooney and M. A. Quaseem, *Chem. Commun.* p. 230 (1967).
5. J. Brault and J. M. Lalancette, *Can. J. Chem.* **42**, 2903 (1964).
6. A. Pelter, T. Levitt, and K. Smith, *Chem. Commun.* p. 435 (1969); J. M. Lalancette, F. Bessette, and J. M. Cliche, *Can. J. Chem.* **44**, 1577 (1966).
7. B. M. Mikhailov and N. S. Fedotoy, *Izv. Akad. Nauk SSSR, Otd. Khim. Nauk* p. 999 (1961).
8. J. M. Lalancette and A. Lachance, *Can. J. Chem.* **47**, 859 (1969).
9. R. H. Cragg, J. P. N. Husband, and A. F. Weston, *Chem. Commun.* p. 1701 (1970).
10. D. J. Pasto, *J. Am. Chem. Soc.* **84**, 3777 (1962); D. J. Pasto, C. C. Combo, and J. Fraser, *ibid.* **88**, 2194 (1966).
11. A. Finch and J. Pearn, *Tetrahedron* **20**, 173 (1964).
12. A. Pelter, K. Rowe, D. N. Sharrocks, and K. Smith, *Chem. Commun.* p. 531 (1975).
13. H. C. Brown and B. C. SubbaRao, *J. Am. Chem. Soc.* **78**, 5964 (1956).
14. J. R. Johnson and M. G. Van Campen, Jr., *J. Am. Chem. Soc.* **60**, 121 (1938).
15. H. C. Brown, U.S. Patent 3,439,046 (1969); *Chem. Abstr.* **71**, 50273c (1969).
16. R. Köster and Y. Morita, *Angew. Chem., Int. Ed. Engl.* **5**, 580 (1966).
17. G. W. Kabalka and H. C. Hedgecock, Jr., *J. Org. Chem.* **40**, 1776 (1975).
18. H. G. Kuivila and E. K. Easterbrook, *J. Am. Chem. Soc.* **73**, 4629 (1951).
19. R. L. Bruce and A. A. Humffray, *Aust. J. Chem.* **24**, 1085 (1971).
20. H. C. Brown, W. R. Heydkamp, E. Breuer, and W. S. Murphy, *J. Am. Chem. Soc.* **86**, 3565 (1964).
21. J. G. Sharefkin and H. D. Banks, *J. Org. Chem.* **30**, 4313 (1965).
22. J. J. Tufariello, P. Wojtkowski, and L. T. C. Lee, *Chem. Commun.* p. 505 (1967).
23. J. J. Tufariello and L. T. C. Lee, *J. Am. Chem. Soc.* **88**, 4757 (1966).
24. G. Köbrich and H. Merkle, *Chem. Ber.* **100**, 3371 (1967).
25. A. Suzuki, S. Nozawa, N. Miyaura, M. Itoh, and H. C. Brown, *Tetrahedron Lett.* p. 2955 (1969).
26. D. J. Pasto and P. W. Wojtkowski, *Tetrahedron Lett.* p. 215 (1970).
27. J. Hooz and S. Linke, *J. Am. Chem. Soc.* **90**, 5936 (1968).
28. J. Hooz and S. Linke, *J. Org. Chem.* **36**, 1790 (1970).
29. A. Suzuki, S. Sono, M. Itoh, H. C. Brown, and M. M. Midland, *J. Am. Chem. Soc.* **93**, 4329 (1971).
30. H. C. Brown, M. M. Midland, and A. B. Levy, *J. Am. Chem. Soc.* **94**, 2114 (1972).
31. H. C. Brown, M. M. Rogić, and M. W. Rathke, *J. Am. Chem. Soc.* **90**, 6218 (1968).
32. H. C. Brown, M. M. Rogić, M. W. Rathke, and G. W. Kabalka, *J. Am. Chem. Soc.* **90**, 818 (1968).
33. B. M. Mikhailov, Yu. N. Bubnov, S. A. Korobeinikova, and S. I. Frolov, *J. Organomet. Chem.* **27**, 16S (1971).
34. H. C. Brown, "Hydroboration," Benjamin, New York, 1962.
35. D. Devaprabhakara and P. D. Gardner, *J. Am. Chem. Soc.* **85**, 1458 (1963).
36. H. C. Brown and R. L. Sharp, *J. Am. Chem. Soc.* **88**, 5851 (1966).
37. T. A. Woods, T. E. Boyd, E. R. Biehl, and P. C. Reeves, *J. Org. Chem.* **40**, 2416 (1975).
38. H. C. Brown and R. L. Sharp, *J. Am. Chem. Soc.* **90**, 2915 (1968).
39. D. J. Pasto and R. Snyder, *J. Org. Chem.* **31**, 2773 (1966).
40. I. J. Borowitz and G. J. Williams, *J. Org. Chem.* **32**, 4157 (1967).

41. G. Zweifel and J. Plamondon, *J. Org. Chem.* **35**, 898 (1970).
42. H. C. Brown and N. Ravindran, *J. Org. Chem.* **38**, 182 (1973).
43. G. Zweifel and H. C. Brown, *J. Am. Chem. Soc.* **86**, 393 (1964).
44. U. E. Diner, M. Worsley, J. W. Lown, and J. A. Forsythe, *Tetrahedron Lett.* p. 3145 (1972).
45. G. W. Kabalka and J. D. Baker, *J. Org. Chem.* **40**, 1834 (1975).
46. W. Siebert, F. R. Rittig, and M. Schmidt, *J. Organomet. Chem.* **25**, 305 (1970).
47. W. Siebert, *Chem. Ber.* **103**, 2308 (1970).
48. E. Negishi and R. E. Merrill, *Chem. Commun.* p. 860 (1974).
49. E. J. Corey and D. J. Beames, *J. Am. Chem. Soc.* **95**, 5829 (1973).
50. E. J. Corey and A. P. Kozikowski, *Tetrahedron Lett.* p. 925 (1975).
51. M. Vidal and P. Arnaud, *Bull. Soc. Chim. Fr.* p. 675 (1972).
52. G. Boireau, D. Abenhaim, C. Bernardon, E. Henry-Basch, and B. Sabourault, *Tetrahedron Lett.* p. 2521 (1975).
53. A. Yasuda, S. Tanaka, K. Oshima, H. Yamamoto, and H. Nozaki, *J. Am. Chem. Soc.* **96**, 6513 (1974).

11

Other Applications of the HSAB Principle

A number of HSAB-pertinent organochemical subjects which have not been discussed in the previous chapters will now be considered. Again the selection of topics is arbitrary, illustrative rather than comprehensive. Some notable exceptions to the HSAB principle are also presented.

11.1. SOLUBILITY AND PROTONATION

The saying that like things dissolve each other can be translated in HSAB terms as hard solvents dissolve hard solutes and soft solutes dissolve in soft solvents.

Water is a typical hard solvent which solubilizes hard acids, hard bases, and hard complexes. The electromotive series indicates that Pt, Hg, Au, Cu, Ag, and the like are unreactive and all these metals form soft ions. Their softness has a direct bearing on their low reactivity in an aqueous environment.

A series of well-defined macrocyclic polyethers (crown ethers) have been synthesized by Pedersen (1, 2), who also demonstrated the capability of complexing and lipophilizing alkali metal ions. The $KMnO_4$-18-crown-6 complex (1) is soluble in benzene (3). Potassium metal is solubilized by a dinaphthalenated cyclic polyether to form the interesting $K^\oplus$-encapsulated anion radical (2) (4).

Guanidinium ion acts as a guest in the ethereal cavity during cyclization of compound 3. A template effect is achieved through specific hard–hard hydrogen bondings (5).

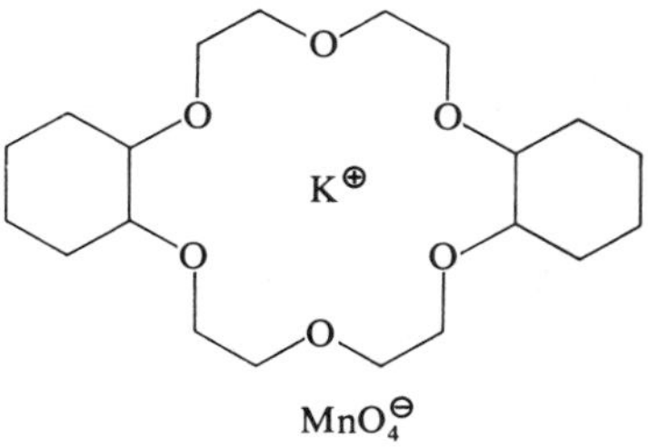

(1)

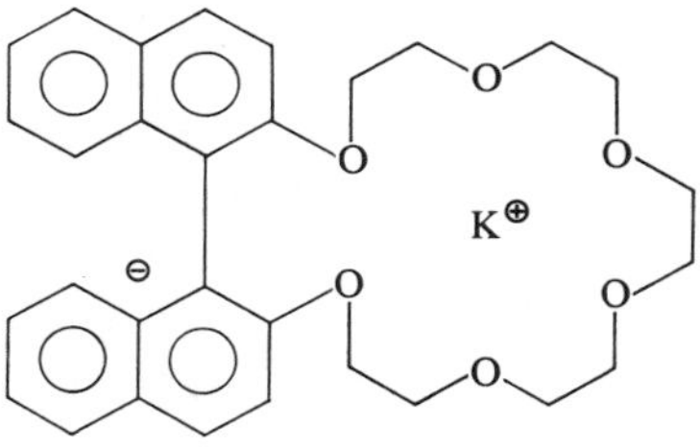

(2)

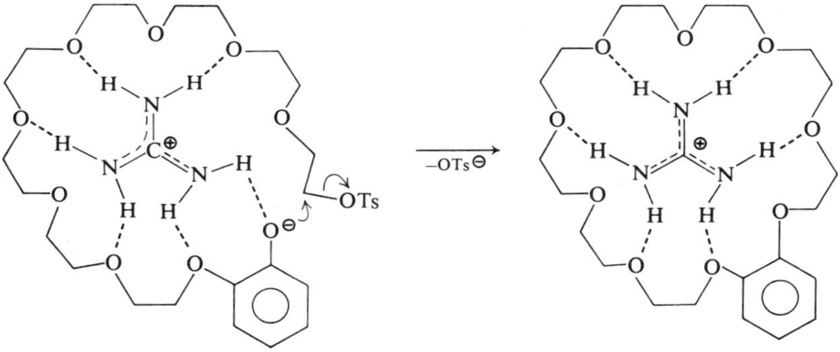

(3)

All the polyethers provide specific residence sites for hard metal ions of appropriate radii; cyclic polysulfides of similar structures prefer complexing soft heavy metal ions (6).

Neutralization of enolates such as that of acetoacetic ester gives the enols instead of the thermodynamically more stable ketones directly. Protonation at the hard basic site agrees with the HSAB principle. Initial protonation of nitronate ions occurs at oxygen also.

The nonconjugated diene is produced on protonation of pentadienyl anion (7). The central carbon atom is harder.

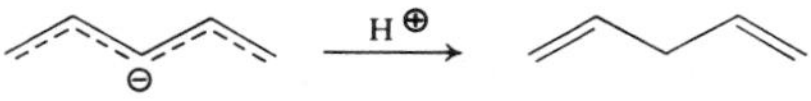

Metallocenes can be protonated at the metal site (8) by strong acids. The M–$H^{\oplus}$ bond formation is more facile when the metal is harder, e.g., Fe > Ru > Os. Acylferrocenes undergo protonation (9) only at the carbonyl oxygen under stable ion conditions. The oxygen atom in these molecules is by far the harder base compared to iron.

11.2. CARBENES AND NITRENES

Thermolysis of phenyl trihalomethylmercurys gives dihalocarbenes. This reaction works well when one of the halogens is bromine, or better, iodine. For the iodo derivatives, decomposition occurs at room temperature (10). The decomposition of phenyl trichloromethylmercury is best accomplished by the addition of iodide ion (11) which has a high affinity for mercury ion because both are soft.

$$\text{PhHg} \overset{\frown}{\underset{\underset{Y}{|}}{}} \text{CX}_2 \longrightarrow \ :\text{CX}_2 + \text{PhHgY}$$

$$Y = \text{Br, I}$$

$$\text{PhHgCCl}_3 + \text{I}^{\ominus} \longrightarrow \text{PhHgI} + \text{CCl}_3^{\ominus}$$

$$\text{CCl}_3^{\ominus} \longrightarrow \ :\text{CCl}_2 + \text{Cl}^{\ominus}$$

An allylic halogen atom is capable of directing the attack of carbenes (carbenoids) by acting as a temporary trap for these species (12). Bromine is more efficient than chlorine, as expected.

$$\text{MeCH=CHCH}_2\text{X} + :\text{C(COOMe)}_2 \longrightarrow \overset{\text{MeCH=CH}}{\underset{\text{(MeOOC)}_2\overset{\ominus}{\text{C}}-\text{X}\overset{\oplus}{}}{}}\text{CH}_2 \longrightarrow \overset{\text{MeCH--CH=CH}_2}{\underset{\text{XC(COOMe)}_2}{|}}$$

Although allyl ethers have a hard oxygen atom they do not react in the same way. It should be noted that complexation of such oxygen with the harder bis-(methoxycarbonyl)carbenoid is more facile than with methylenoid.

The Reimer–Tiemann reaction probably involves the attack of dichloro-carbene at the carbon of the ambient phenoxide ion. Conversely, only the oxy site of phenoxide ion intercepts the much harder difluorocarbene (13). Dibromocarbene attacks on the soft donor atom of thiophenoxide to give triphenylthiomethane (14) via further S_N2 displacement of the intermediate.

$$\text{PhSNa} + :\text{CBr}_2 \xrightarrow{\text{ROH}} \text{PhSCHBr}_2 \xrightarrow{\text{PhS}^{\ominus}} (\text{PhS})_3\text{CH}$$

Although the trapping of carbenes by hard bases is unfavorable, as further substantiated by the nontransient survival of dihalocarbenes in strong alkaline media, circumstances may indicate such a route as the only available pathway. Thus, azodiformate esters react with dichlorocarbene via their nitrogen atom (15).

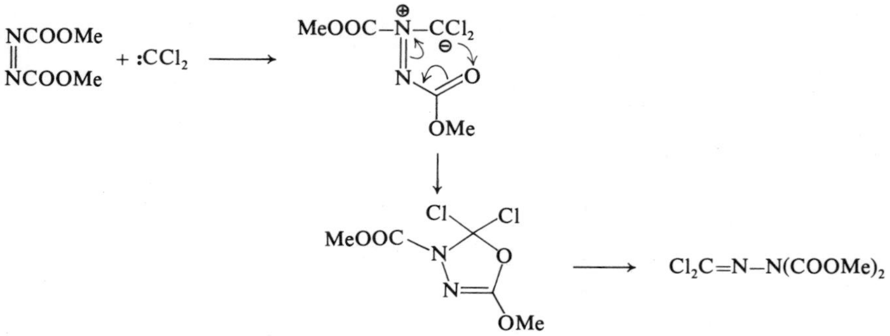

$$NCOOMe \atop \| \quad + :CCl_2 \quad \longrightarrow \quad MeOOC\overset{\oplus}{-N-CCl_2} \atop NCOOMe$$

(reaction scheme)

$$\longrightarrow \quad Cl_2C=N-N(COOMe)_2$$

Diazo compounds having a β heterosubstituent decompose via carbenic rearrangement to afford olefinic products (16). Only hydrogen shift occurs in cases where the substituents are OR or NMe$_2$. Thioether rearrangement predominates when the β group is SR. The implication from these different results is clear. Whereas hard basic functions do not interact with the neighboring carbenic center, specific soft interactions are responsible for the shift of the sulfur substituent.

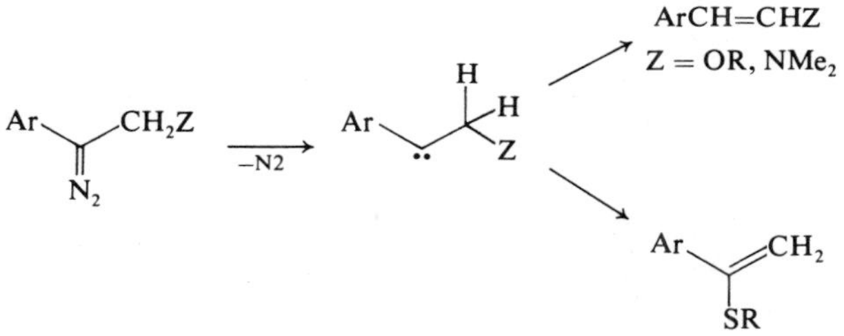

(reaction scheme)

ArCH=CHZ
Z = OR, NMe$_2$

Styrylidine generated by thermolysis of the benzylidene derivative of Meldrum's acid undergoes internal hydride migration three times as fast as the phenyl shift (17).

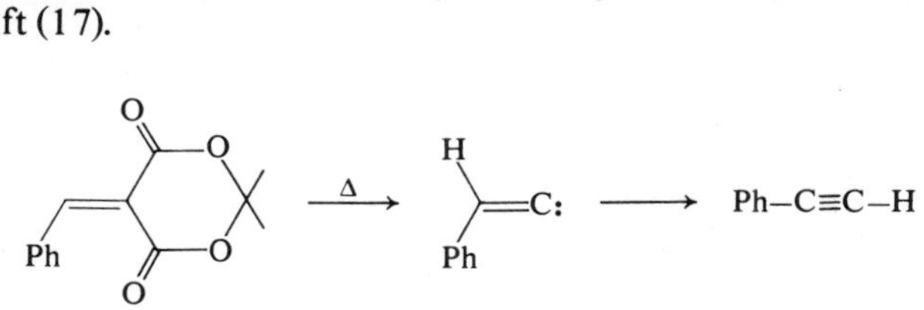

(reaction scheme)

$$Ph-C{\equiv}C-H$$

The migratory aptitudes toward carbenes (16, 18, 19), H > Ph > Me and SH > H > OR,NR$_2$, parallel the relative softness of the groups. It is important to note that π participation is involved during phenyl shift. The mechanism by which a neighboring hydrogen migrates to a carbenic center has been investigated by *ab initio* molecular orbital calculations (20).

1-Trimethylsilyl-1-trimethylsiloxy-2-methylpropene undergoes thermal decomposition to the unsaturated carbene and hexamethyldisiloxane (21). The driving force for this reaction is the favorable Si–O bond formation between the hard acid and base moieties.

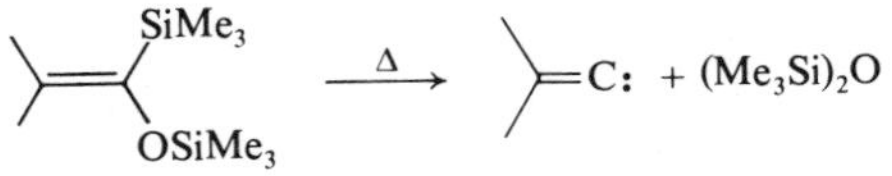

Metal–carbene complexes react with Wittig reagents to give olefins (22). The initial C—C bond formation is a soft–soft interaction; subsequent decomposition of the intermediates is also highly favored as the metal complexes generated are symbiotically stabilized.

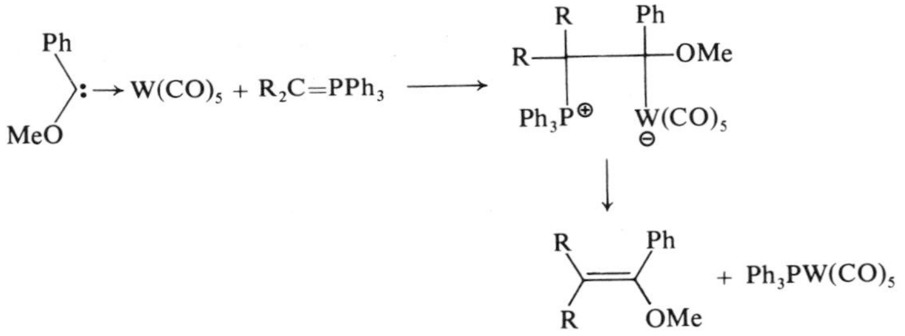

Nitrenes are soft acceptors also. Arylnitrenes from azide decomposition are readily intercepted by carbon monoxide which is a soft donor (23). The reduction of nitroarenes (24, 25) by CO also involves soft interactions.

$$ArNO_2 \xrightarrow{CO} ArN(O^{\ominus})\text{–}O\text{–}C{\equiv}O^{\oplus} \xrightarrow[-CO_2]{} ArN{=}O$$

$$ArN{=}O \xrightarrow{CO} ArN^{\ominus}\text{–}O\text{–}C{\equiv}O^{\oplus} \xrightarrow[-CO_2]{} ArN\!: \xrightarrow{CO} ArN{=}C{=}O$$

Isonitriles are generally considered hybrids having the canonical structures

$$R\overset{\oplus}{N}{\equiv}C^{\ominus} \longleftrightarrow RN{=}C\!:$$

They exhibit typical soft characteristics in reactions (26) with Michael acceptors, positive halogen compounds, and elemental sulfur. They intercept carbenes (27–30), nitrenes (31), and benzyne (32) to give α-addition products.

$$RNC + CH_2{=}CHCOOMe \longrightarrow RN{=}\overset{\oplus}{C}CH_2\overset{\ominus}{C}HCOOMe \xrightarrow{\overset{\ominus}{H}\sim} RN{=}CHCH{=}CHCOOMe$$

Cycloaddition of isonitrile to cycloalkynes (33) is also typical of carbene reactions.

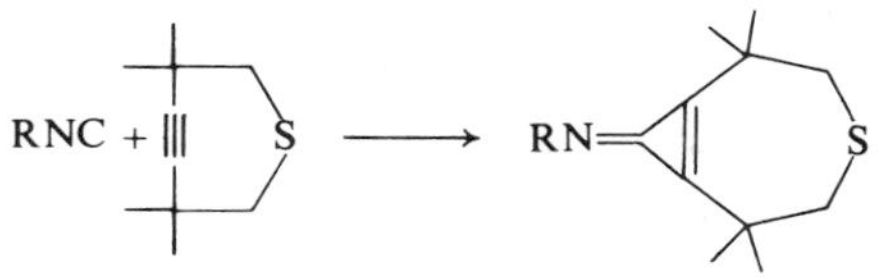

Conversion of isonitriles to carbamates by thallium(III) nitrate in the presence of an alcohol (34) is initiated by a soft–soft interaction between C and Tl atoms.

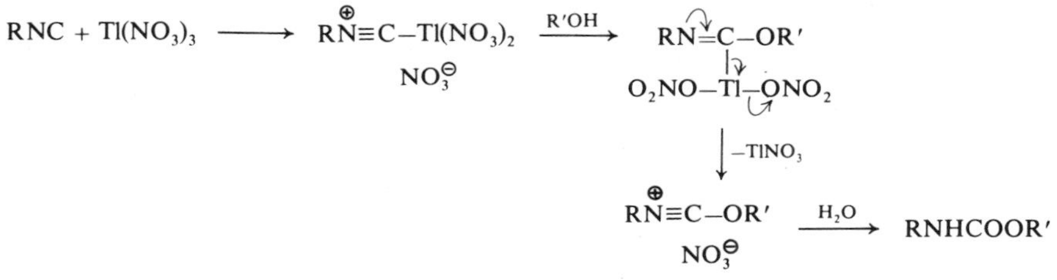

Aminometallation (35) is also observed when isonitriles are treated with organolead and organotin amino compounds. This is a facile biphilic process.

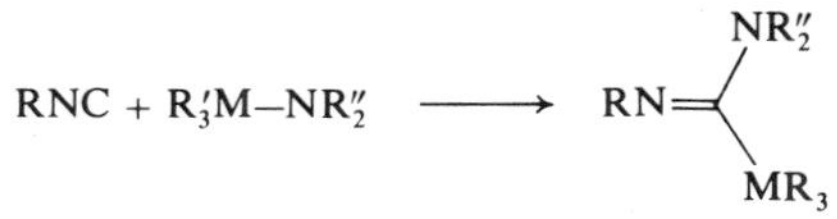

The contribution of the carbene-like structure to isonitrile molecules renders them electrophilic as evidenced by condensation with active methylene compounds (36a) and phenols (36b).

11.3. ORGANIC CHEMISTRY OF GROUP IV ELEMENTS

A neutral silicon center is invariably a harder acceptor than the carbon counterpart. However, it is interesting to note that the direction of attack on Si of optically active 2-(α-naphthyl)-2-phenyloxasilacyclopentane by organometallic reagents is dictated by the hardness of the nucleophile (37). Harder bases interact with the vacant d_{xy} orbital of Si, whereas softer bases attack axially on the d_{z^2} of silicon.

Hard bases attack silacyclobutanes at the heteroatom exclusively, whereas soft nucleophiles such as the Wittig reagents cause rupture of these molecules at a C—C bond (38).

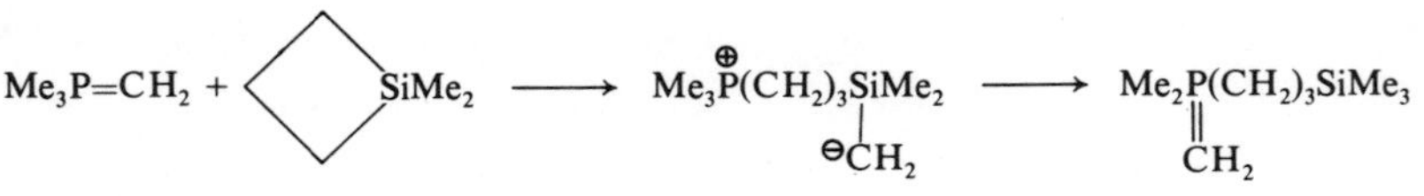

$$Me_3P=CH_2 + \underset{SiMe_2}{\diamond} \longrightarrow \underset{\ominus CH_2}{Me_3\overset{\oplus}{P}(CH_2)_3SiMe_2} \longrightarrow Me_2\underset{\parallel}{P}(CH_2)_3SiMe_3 \;\; \underset{CH_2}{}$$

The unusual fragmentation of a disilane (39) promoted by ethoxide ion as illustrated in the following reaction must be caused in part by the high affinity of Si for hard bases.

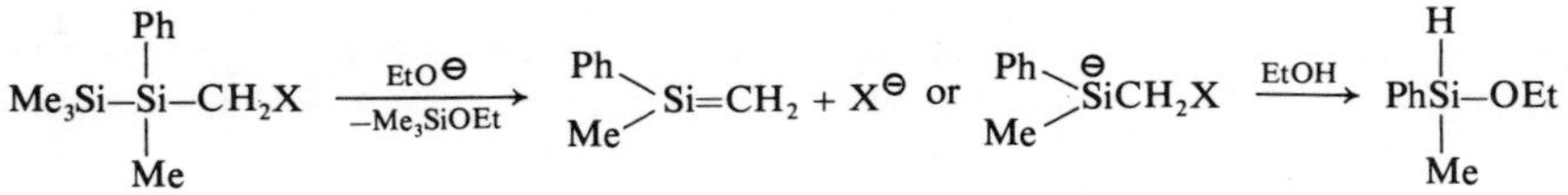

$$\underset{Me}{\overset{Ph}{Me_3Si-Si-CH_2X}} \xrightarrow[-Me_3SiOEt]{EtO^{\ominus}} \underset{Me}{\overset{Ph}{>}}Si=CH_2 + X^{\ominus} \;\text{or}\; \underset{Me}{\overset{Ph}{>}}\overset{\ominus}{Si}CH_2X \xrightarrow{EtOH} \underset{Me}{\overset{H}{PhSi-OEt}}$$

Another unexpected reaction pattern emerges from the acetolysis of dimethylsilacyclohexan-4-ol tosylate (40). The dominating factor is the hardness of the Si atom.

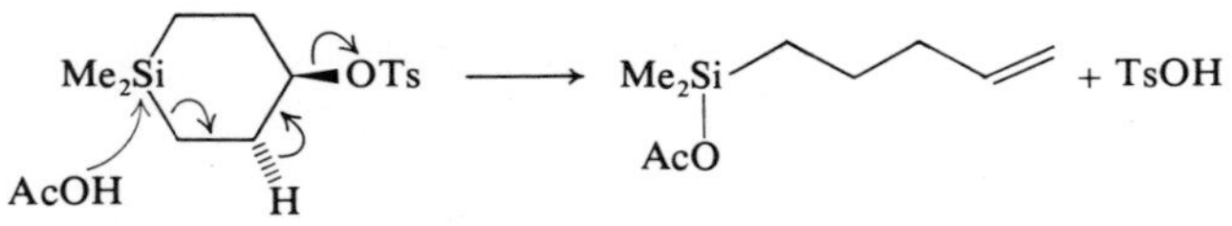

$$\underset{AcOH}{Me_2Si}\big\langle\big\rangle\text{-OTs} \longrightarrow \underset{AcO}{Me_2Si}\diagup\diagdown\diagup\diagdown\!\!= + \text{TsOH}$$

The thermal silicon–Pummerer rearrangement (41, 42) leads to a more compatible partnership among the various hard and soft moieties of the molecules.

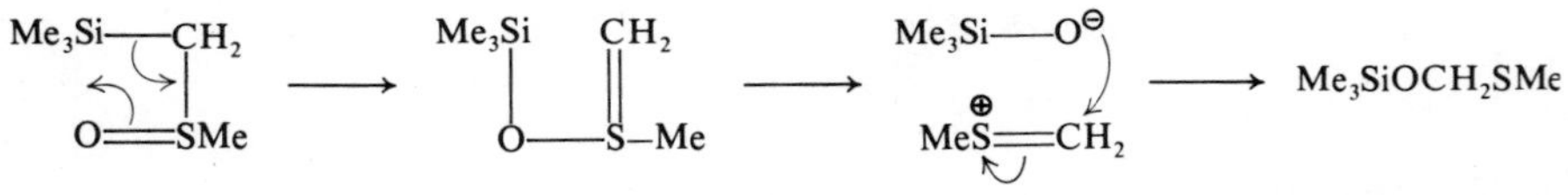

$$\underset{O=\!=SMe}{Me_3Si-CH_2} \longrightarrow \underset{O-\!-S-Me}{Me_3Si \;\; CH_2} \longrightarrow \underset{MeS=\!\!=\overset{\oplus}{}CH_2}{Me_3Si-\!-O^{\ominus}} \longrightarrow Me_3SiOCH_2SMe$$

It has been postulated that intermediates containing Si–Tl bond are formed during the oxidation of aryldimethylsilanes (43) with thallium triacetate. The silicon acts as a nucleophile toward Tl(III). However, the following mechanism consisting of a HSAB-matched transition state appears to be a better description.

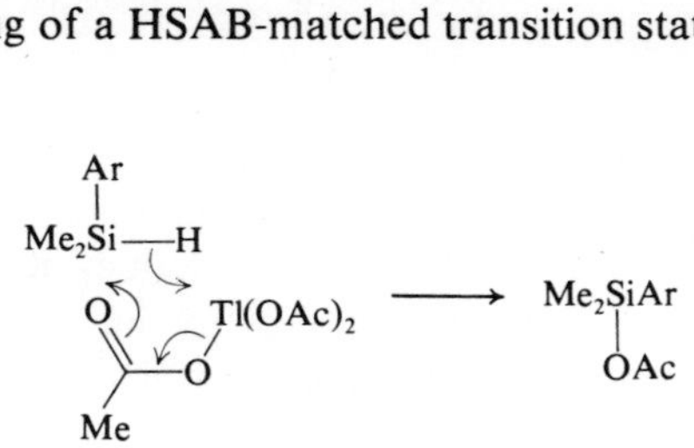

$$\underset{Me}{\overset{Ar}{\underset{O}{Me_2Si-H}}}\;Tl(OAc)_2 \longrightarrow \underset{OAc}{Me_2SiAr}$$

The observations (44) that nucleophilic attack on acyloxysilanes by organometallics occurs at carbonyl and that the hard bases (ROH, RO$^{\ominus}$, RNH$_2$)

bind preferentially to silicon are by no means at variance with the HSAB theory, as silicon is harder than carbonyl carbon.

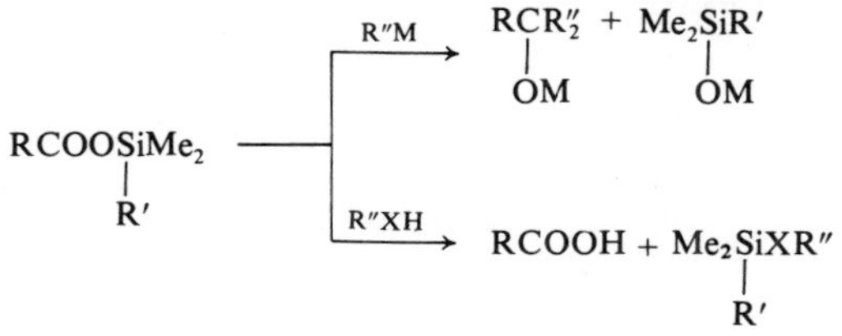

$$
\begin{array}{c}
\text{RCOOSiMe}_2 \\
\quad\mid \\
\quad R'
\end{array}
\xrightarrow{\substack{R''M \\ \\ R''XH}}
\begin{array}{c}
\underset{\text{OM}}{\text{RCR}''_2} + \underset{\text{OM}}{\text{Me}_2\text{SiR}'} \\[2ex]
\underset{\quad R'}{\text{RCOOH} + \text{Me}_2\text{SiXR}''}
\end{array}
$$

The quenching of dihydropyran anion with trimethylsilyl chloride yields only 2-trimethylsilyl-Δ^3-dihydropyran (45). The reaction takes place at the harder of the two basic centers (C-2, C-4). Carbon-2 is the harder because it is flanked on one side by an oxygen atom.

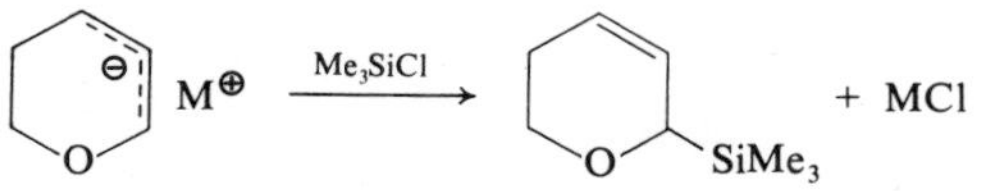

The pseudonitrosite of ω-styryltrimethylsilane is decomposed by alcoholic hydroxide to furnish β-nitrostyrene (46). Silicophilic attack is preferred to abstraction of the very active hydrogen alpha to the nitro group.

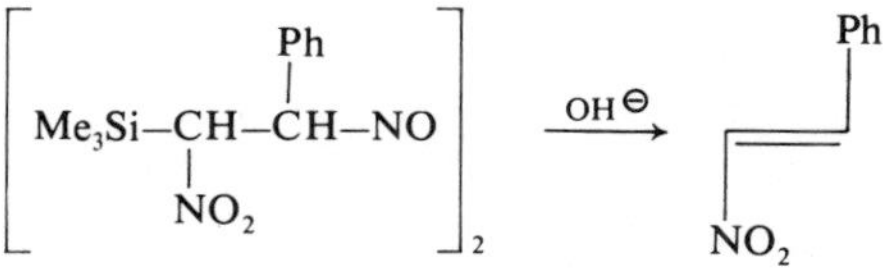

α-Substituted benzyltriphenylsilanes rearrange on heating to products in which the harder groups are silicon-bound (47), apparently due to symbiosis. The base-catalyzed and thermal rearrangements of α-silylacetic acids (48) and the conversion of β-ketosilanes to siloxyalkenes (49) operate on the same principle.

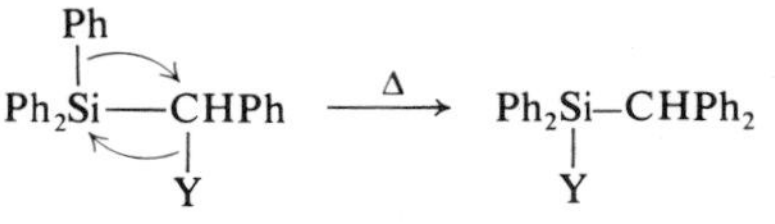

$$
\begin{array}{c}
\overset{\displaystyle \text{Ph}}{\underset{}{\mid}} \\
\text{Ph}_2\text{Si}\!-\!\text{CHPh} \\
\mid \\
\text{Y}
\end{array}
\xrightarrow{\ \Delta\ }
\begin{array}{c}
\text{Ph}_2\text{Si}\!-\!\text{CHPh}_2 \\
\mid \\
\text{Y}
\end{array}
$$

$$\text{Y} = \text{F, Cl, OAc, OTs}$$

$$
\begin{array}{c}
\text{R}' \\
\mid \\
\text{R}_3\text{Si}\!-\!\text{C}\!-\!\text{COOH} \\
\mid \\
\text{R}''
\end{array}
\longrightarrow
\begin{array}{c}
\text{R}' \\
\mid \\
\text{H}\!-\!\text{C}\!-\!\text{COOSiR}_3 \\
\mid \\
\text{R}''
\end{array}
$$

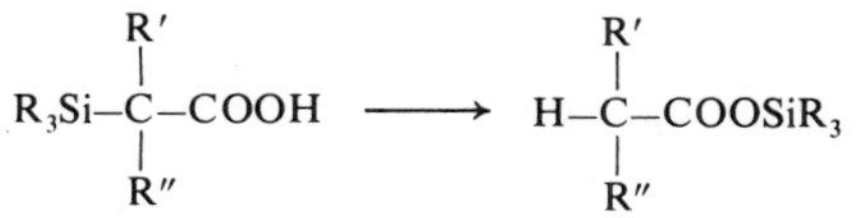

Although the dyotropic rearrangement of α-siloxyalkylsilanes (50a) proceeds readily at moderate temperatures, decomposition is the major course for the sulfur analogs (50b).

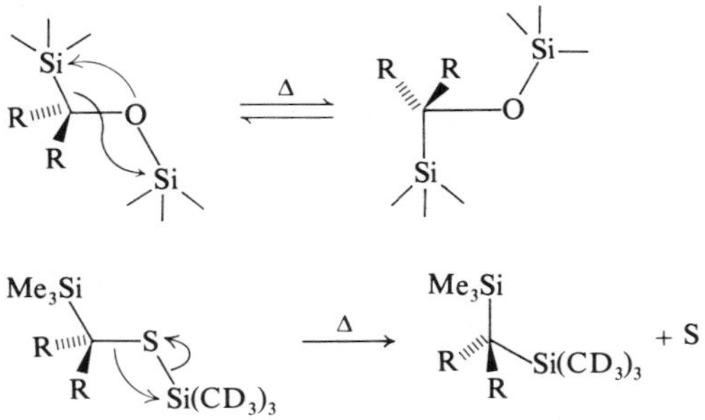

Trimethylsilylmethylenephosphoranes have two different electrophilic sites (P, Si). Reaction of these species with methanol generates methoxytrimethylsilane and methylenephosphoranes (51). Apparently silicon is a harder acceptor than tetrahedral phosphorus.

Desilylation of organosilanes is most efficiently achieved by hard bases such as fluoride ion (52) and oxy bases (53).

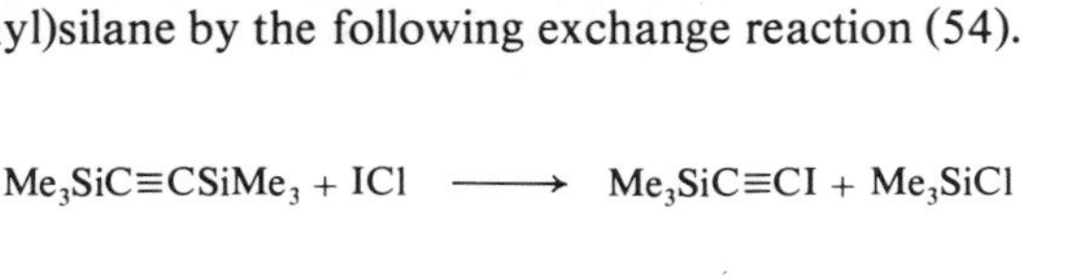

The greater affinity of silicon for hard bases allows the preparation of iodoethynyl(trimethyl)silane by the following exchange reaction (54).

$$Me_3SiC\equiv CSiMe_3 + ICl \longrightarrow Me_3SiC\equiv CI + Me_3SiCl$$

Silylphosphines add onto the C=N group of aldimines to provide phosphinylated silylamines (55). Better matching for both the hard Si and the soft P atoms results.

The same principle underlies the following facile transformation (56).

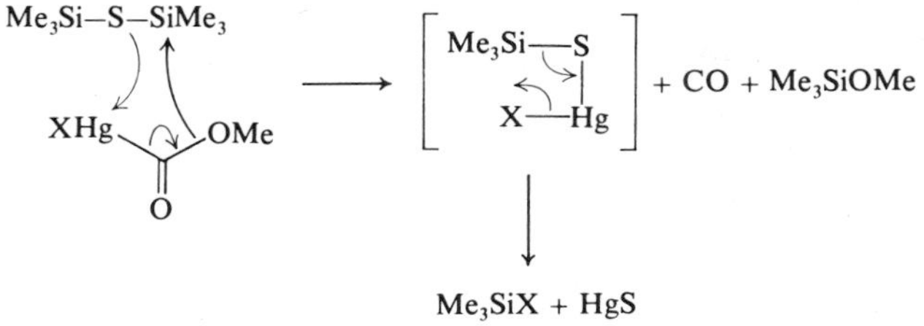

$$Me_3SiX + HgS$$

A method for hydrolysis of esters under neutral conditions has been developed (57a) based on HSAB considerations.

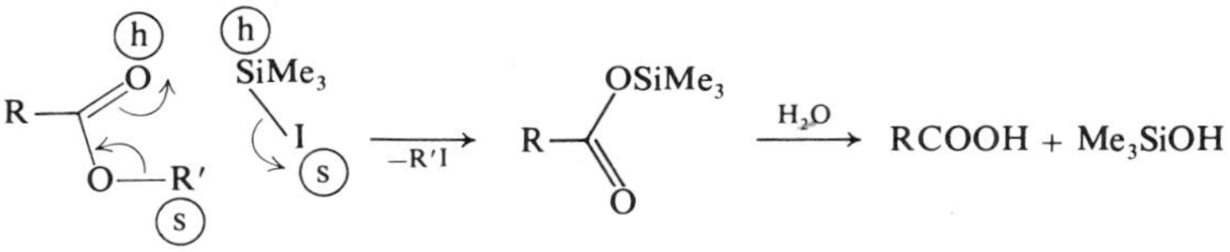

A combination of phenyltrimethylsilane and iodine is even more effective—presumably a favorable termolecular transition state is attained during the reaction (57b).

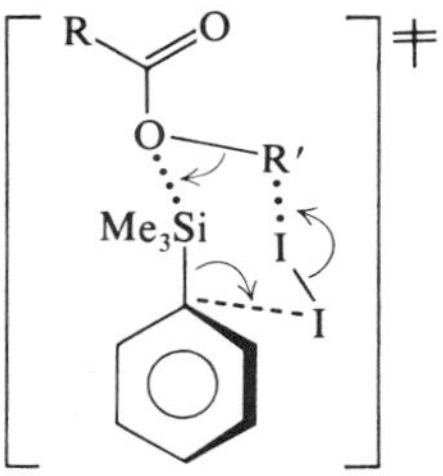

Such a mechanism receives support from the observation that aryl alkyl ethers are readily cleaved by the $PhSiMe_3$-I_2 reagent but only inefficiently by iodotrimethylsilane.

Sulfenamides are formed when sulfenate esters are treated with amines or their trimethylsilyl derivatives (58).

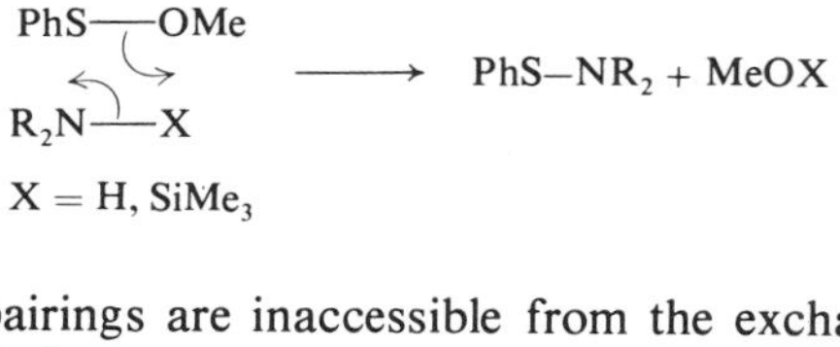

$$PhS-NR_2 + MeOX$$

$$X = H, SiMe_3$$

Although perfect pairings are inaccessible from the exchange, it results in a more compatible relationship.

1,4-Addition to 1,1-bis(trihalomethyl)-2,2-dicyanoethylenes by various metalloid compounds (59) has been reported. Again the HSAB principle is followed.

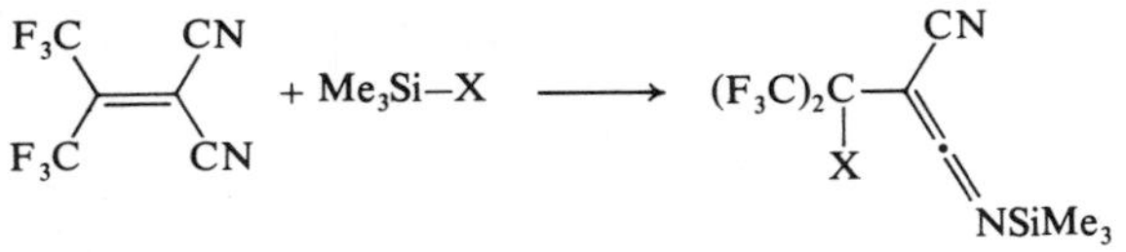

X = H, PhNH, MeS

The trimethylsilyl group of enol silyl ethers can participate in ene reactions with singlet oxygen (60, 61).

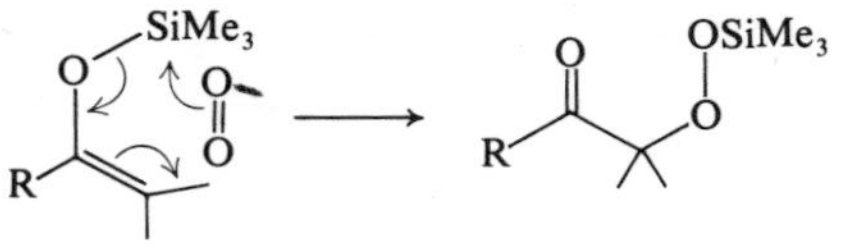

Owing to the hardness of Si(IV) and Sn(IV) nuclei, triorganosilyl and triorganostannyl groups can act as electrofugal moieties during heterolysis initiated by an electron-deficient β atom (62, 63).

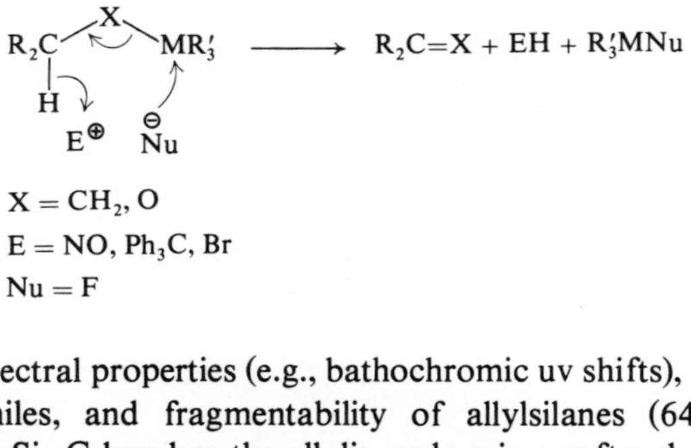

$$R_2C=X + EH + R_3'MNu$$

X = CH_2, O
E = NO, Ph_3C, Br
Nu = F

The unusual spectral properties (e.g., bathochromic uv shifts), high reactivity toward electrophiles, and fragmentability of allylsilanes (64) are due to weakening of the Si—C bond as the allylic carbon is a softer donor compared to a nonconjugated atom.

4-Tributylstannylbut-1-ene undergoes a remarkable cyclodestannylation on exposure to electrophiles (65). Because tin is harder than carbon, the observed fragmentation transfers the net charge to the tin and is energetically profitable.

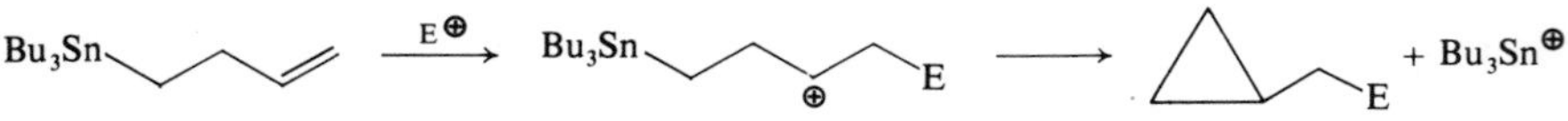

E = I, Br, Cl, HgCl, etc.

Organotin hydrides are synthesized by the reduction of the halides or oxides with lithium aluminum hydride (66). The tin atom of these halides and oxides is softened to a large extent by the organic ligands; therefore, it is a better partner for the hydride ion than aluminum.

In the synthesis of orthocarbonates by the reaction of carbon disulfide with dialkoxystannanes (67, 68), each successive stage is aided by symbiosis.

Lithiotrimethylstannane reacts with cis-4-tert-butylcyclohexyl bromide and the trans-tosylate to yield the same cis product (69). The change in mechanism is due to hard/soft disparity of the leaving groups. The soft bromine can stabilize the transition state leading to a product having a retained configuration because of coordination with Sn.

11.4. REACTIONS OF ORGANOHALIDES, GRIGNARD AND RELATED REAGENTS

The $C-X$ bond (where $X = Br, I$) can be considered as a soft acid–base pair. The halogen may act as either an acid or a base, depending on its environment and on the nature of the reaction involving the breakage of $C-X$ linkage.

Ordinary alkyl halides are thermally stable. On the other hand, cyclopropyl halides may undergo ionization concomitant with fission of a ring $C-C$ bond. The ease of $C-X$ bond cleavage parallels the softness of X. For example, the dibromocarbene adduct of norbornene (70) rearranges during its preparation and the corresponding dichlorocarbene adduct can be isolated. Rearrangement of the latter compound can be brought about by mild heating, whereas a higher temperature is required for breaking the $C-F$ bond in an analogous system (71). Intrinsic bond strength is the determining factor; however, it is believed that softness also contributes to the observed variation.

In the formation of exocyclic bonds cyclopropane utilizes harder orbitals (more s character). It follows that a soft substituent destabilizes the molecule, thus the $C-Br$ bond of bromocyclopropane should be weaker than that of 2-bromopropane. Conversely, the $C-F$ bond of fluorocyclopropane is expected to be as strong, if not stronger, than that of 2-fluoropropane.

The substitution of perfluoroalkyl iodide by methanethiolate ion (72) is not a straightforward S_N2 process. The mechanism has been established as shown.

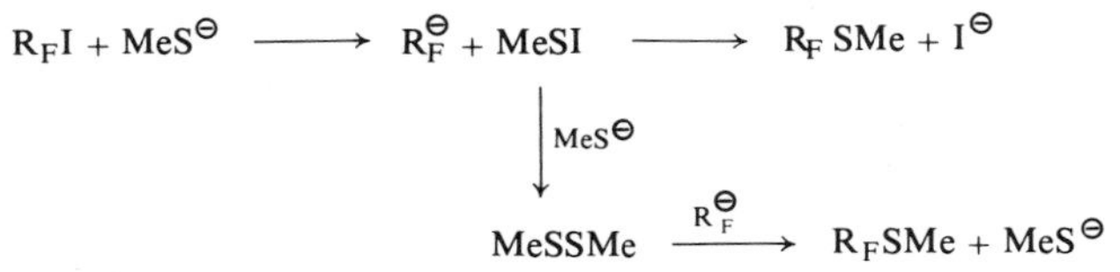

$$R_FI + MeS^\ominus \longrightarrow R_F^\ominus + MeSI \longrightarrow R_F SMe + I^\ominus$$

$$\downarrow MeS^\ominus$$

$$MeSSMe \xrightarrow{R_F^\ominus} R_FSMe + MeS^\ominus$$

Several factors favor the observed pathways: (1) the iodine atom is a soft acid, (2) the alkyl carbon is hardened by fluorine substituents and less prone to interact with the soft base $MeS^\ominus$, and (3) perfluoroalkide ions are good leaving groups.

The debromination of *gem*-dibromocyclopropanes by dimsyl anion (73) involves removal of a soft halogen by the carbon base. The formation of allenes from the reaction of *gem*-dibromocyclopropanes with alkyllithiums (74) must proceed at its early stages by a similar mechanism. Halogen–metal exchange of a *gem*-dibromocyclopropanecarboxylic acid (75) is 2.5 times faster than proton abstraction from the acid! This is a remarkable demonstration of the HSAB axiom.

1-Chloro-1-(cyclopropylethynyl)cyclopropane undergoes halide–metal exchange with *n*-BuLi. Interestingly, an S_N2 displacement of the chlorine by a phenyl group of the harder PhLi reagent is noted (76).

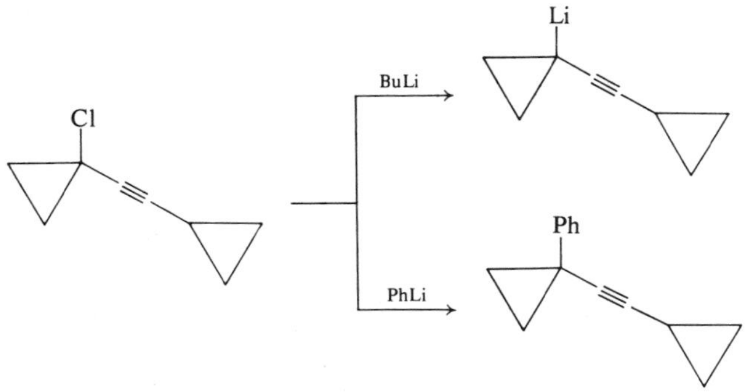

Masamune *et al.* (77) have studied the stereochemical consequences of reductive removal of halo (Cl,Br) and mesyloxy groups with a Cu(I) hydride reagent.

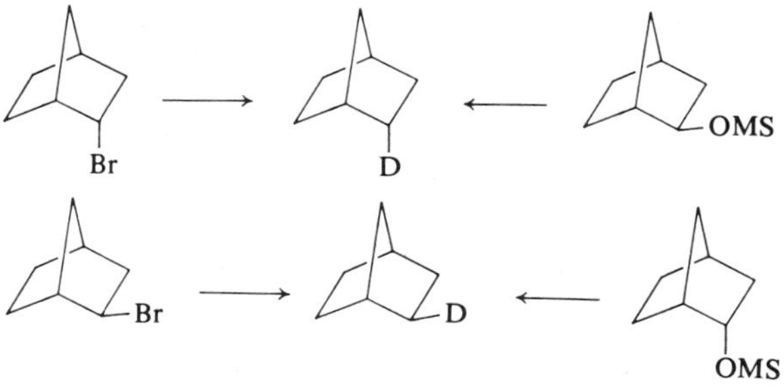

Displacement of the mesyloxy group is formally a S_N2 process. The hydride reaction with the bromo compounds probably involves electron transfer, capture of bromine, and back-donation of hydrogen (deuterium) to the substrates within the ligand sphere of the copper complexes. The reason for the dichotomy must be hinged on the acceptor characteristics of bromine vis-à-vis the harder carbon.

Similar duality exists during the coupling studies of benzyl derivatives with diorganocuprates (78). It is now well known that diorganocuprates are soft nucleophiles which displace organohalides (bromides, iodides) with ease (79, 80).

$$PhCH_2\overset{\oplus}{N}Me_3 \ I^{\ominus} \xrightarrow{\ n\text{-}Bu_2CuLi\ } PhCH_2Bu$$

$$2PhCH_2X \xrightarrow{\ R_2CuLi\ } PhCH_2CH_2Ph$$

$$X = halide$$

Interaction of 1-phenyl-6-halohexyne with lithium dialkylcuprates (81) leads to alkylated derivatives and benzylidenecyclopentane. The proportion of the cyclized product increases if either the halogen of the acetylene or R_2CuLi becomes softer.

The nucleophilic displacement of 1-aryl-2-haloacetylenes with various bases (82) correlates quite well with the HSAB principle with respect to the attacking sites.

Both bromine atoms at C-4 of 2,4,4,6-tetrabromocyclohexa-2,5-dienone are extracted by di(phenylethynyl)mercury (83), presumably via the mechanism shown below.

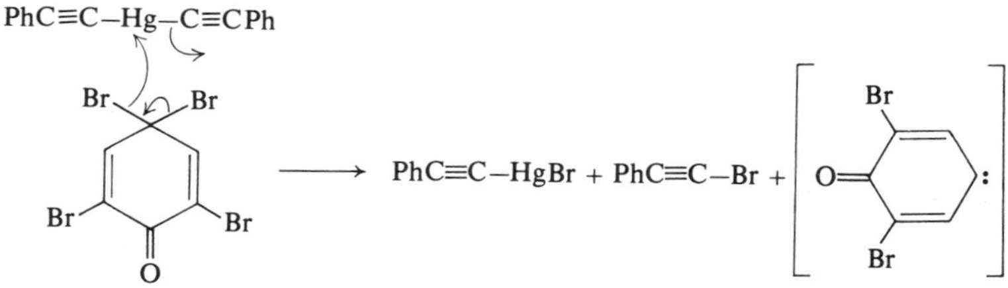

The dienone is a good donor of $Br^{\oplus}$ and it may be used for promoting the dimerization of thiols (84). The efficiency of this process is quite remarkable in comparison with the bromination of harder substrates by the same reagent (85). Cyanogen bromide also effects disulfide formation (86).

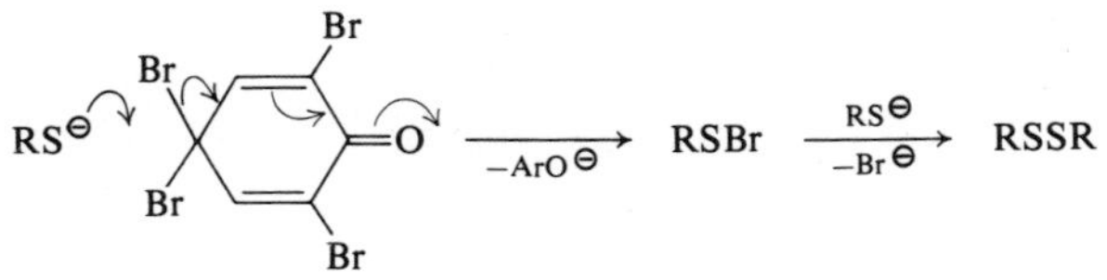

The halogen of halo(triphenyl)phosphonium salts is removed by organometallic agents (87), although substitution at P can intervene when the reaction is carried out at higher temperatures.

It has been observed that organomagnesium fluorides (88) react faster with benzophenone and give a higher ratio of addition to reduction products than the corresponding bromo reagents. Both aspects attest to a harder nature of RMgF as would be expected.

α-Cyanoepoxides react with magnesium compounds in a number of ways (89). Magnesium bromide and Grignard reagents coordinate with the epoxy oxygen with the hard Mg atom prior to nucleophilic attack. Dialkylmagnesium compounds, while having a softer metal center, complex with the nitrile function.

Grignard discovered that alkylmagnesium halides react with cyanogen chloride and cyanogen iodide in different ways (90).

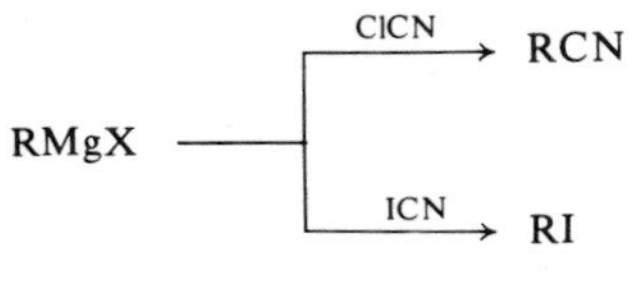

More recently, cleavage of the aryl C–Sn bond by cyanogen halides (91) has been shown to give either aryl bromides or cyanides. The dichotomy may be readily explained by the difference in the softness of the acceptors $I^{\oplus} > Br^{\oplus} > CN^{\oplus} > Cl^{\oplus}$.

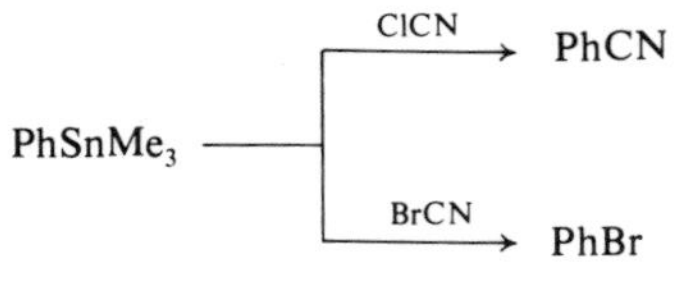

In a synthesis of sulfonylnitriles (92) from sodium sulfinates, the soft sulfur nucleophilic center binds to the cyano group of cyanogen chloride.

Benzoyl fluoride is more reactive than benzoyl chloride toward phenyl-magnesium bromide (93). However, the reactivity of acyl halides with di-*p*-tolylmercury (94) decreases according to the sequence $RCOI > RCOBr > RCOCl > RCOF$.

11.5. EXCEPTIONS

For an empirical principle with its theoretical foundation yet to be firmly and thoroughly established, exceptions to its general validity are bound to exist. In the case of the HSAB principle, there are remarkably few. These exceptions may simply mirror our presently imperfect understanding of certain subtle bonding aspects or other undetermined factors.

Hudson has stressed that many pitfalls lie in the area of ambident reactivity which could lead to interpretive labyrinths. A knowledge of the nature of the transition state is imperative in avoiding these pitfalls.

The apparent contradiction to symbiosis during acetolysis of hex-5-enyl derivatives (95) is inferred from the observation that more double-bond participation products arise from the arenesulfonates than the softer halides. Pearson (96) argues that the solvolysis may be more complex than a simple S_N2 type displacement and as such there is no reason for symbiosis to be the dominant factor.

Exclusive N-alkylation of diazotates (97) with both Meerwein's reagent and alkyl iodides to give *trans*-azoxyalkanes has been reported.

The Claisen condensation of esters normally gives β-keto esters. Acylation of tricarboalkoxymethide ions also leads to reaction at the carbon atom. The aberration from the HSAB principle could be due to intervention of pre-equilibrium and rearrangement, respectively (98).

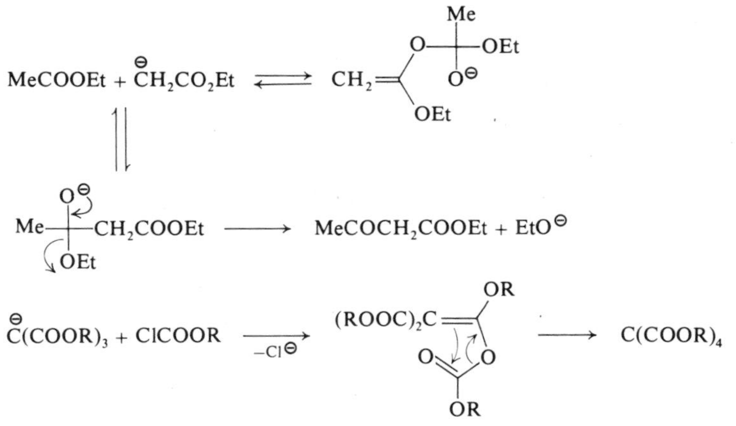

The inter- vs. intramolecular alkylation of 2-bromoethyl ester α anions (99) cannot be explained on simple terms.

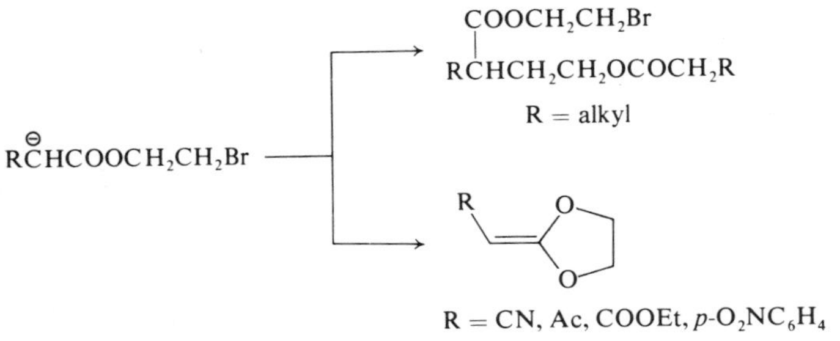

$$R\overset{\ominus}{C}HCOOCH_2CH_2Br$$

$$\begin{array}{c} COOCH_2CH_2Br \\ | \\ RCHCH_2CH_2OCOCH_2R \end{array}$$

R = alkyl

R = CN, Ac, COOEt, p-O$_2$NC$_6$H$_4$

Thallium(I) salts of β-diketones are alkylated exclusively at the central carbon atoms (100). The observed regiospecificity is a consequence of both the softness of the counterion and the heterogeneity of the reaction.

Ether suspensions of these salts undergo O-acetylation on treatment with acetyl chloride at $-78°$, and exclusive C-acylation with acetyl fluoride at ambient temperature. Assuming that acetyl fluoride is the harder acid, the above-mentioned reactions are in violation of the HSAB principle.

Thallio(I)carbazole attacks vinyl acetate (101) to afford N-acetylcarbazole. The potassium salt undergoes α-acetoxyethylation on the nitrogen atom when it is treated with the same reagent. It may be envisaged that during N-acetylation of thalliocarbazole there exists a soft–soft interaction between the double bond of vinyl acetate and the thallium which brings the nitrogen atom and the carbonyl group into juxtaposition and releases electrons for the formation of the N–C bond.

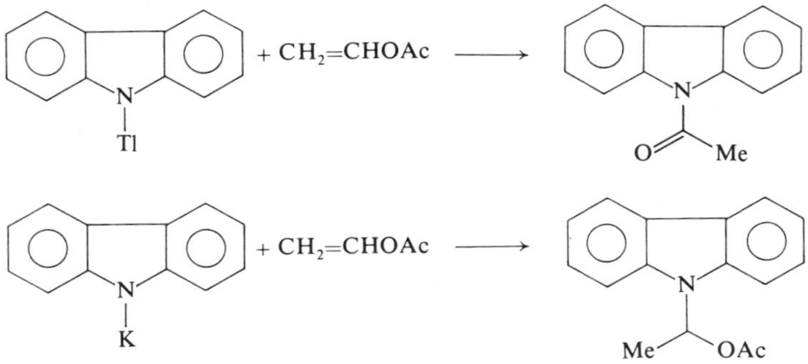

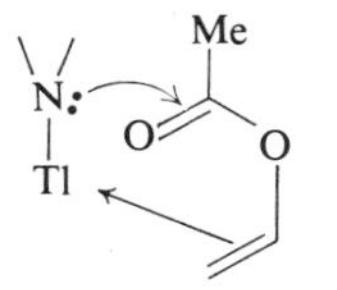

Finally, it should be reiterated that chemical phenomena are often very complicated, and it would be naive to presume that simple concepts such as the HSAB principle can provide answers or rationalizations for each and every observation. Only by sagacious and judicious application can these chemical principles serve the interpreter with reward.

REFERENCES

1. C. J. Pedersen, *J. Am. Chem. Soc.* **89**, 2495 and 7017 (1967).
2. C. J. Pedersen and H. K. Frensdorff, *Angew. Chem., Int. Ed. Engl.* **11**, 16 (1972).
3. D. J. Sam and H. E. Simmons, *J. Am. Chem. Soc.* **94**, 4024 (1972).
4. K. Madan and D. J. Cram, unpublished results, quoted in D. J. Cram and J. M. Cram, *Science* **183**, 803 (1974).
5. K. Madan and D. J. Cram, *Chem. Commun.* p. 427 (1975).
6. K. Travis and D. H. Busch, *Chem. Commun.* p. 1041 (1970).
7. A. G. Catchpole, E. D. Hughes, and C. K. Ingold, *J. Chem. Soc.* p. 8 (1948).
8. T. J. Curphey, J. O. Santer, M. Rosenblum, and J. H. Richards, *J. Am. Chem. Soc.* **82**, 5249 (1960).
9. G. A. Olah and Y. K. Mo, *J. Organomet. Chem.* **60**, 311 (1973).
10. D. Seyferth and C. K. Haas, *J. Org. Chem.* **40**, 1620 (1975).
11. D. Seyferth, J. Y.-P. Mui, M. E. Gordon, and J. M. Burlitch, *J. Am. Chem. Soc.* **87**, 681 (1965).
12. W. Ando, S. Kondo, K. Nakayama, K. Ichibori, H. Kohoda, H. Yamato, I. Imai, S. Nakaido, and T. Migita, *J. Am. Chem. Soc.* **94**, 3870 (1972).
13. T. G. Miller and J. W. Thanassi, *J. Org. Chem.* **25**, 2009 (1960).
14. S. D. Saraf, *J. Nat. Sci. Math.* **11**, 127 (1971); *Chem. Abstr.* **77**, 113948 (1972).
15. D. Seyferth and H. Shih, *J. Am. Chem. Soc.* **94**, 2508 (1972).
16. J. H. Robson and H. Shechter, *J. Am. Chem. Soc.* **89**, 7112 (1967).
17. R. F. C. Brown and K. J. Harrington, *Chem. Commun.* p. 1175 (1972).
18. L. Friedman and H. Shechter, *J. Am. Chem. Soc.* **83**, 3159 (1961); H. Philip and J. Keating, *Tetrahedron Lett.* p. 523 (1961).
19. W. Kirmse and M. Buschkoff, *Chem. Ber.* **100**, 1491 (1967).
20. J. A. Altmann, I. G. Csizmadia, and K. Yates, *J. Am. Chem. Soc.* **97**, 5217 (1975).
21. P. J. Stang and D. P. Fox, *Abstr. 172nd Am. Chem. Soc. Meet., San Francisco,* ORGN 97 (1976).
22. C. P. Casey and T. J. Burkhardt, *J. Am. Chem. Soc.* **94**, 6543 (1972).
23. R. P. Bennett and W. B. Hardy, *J. Am. Chem. Soc.* **90**, 3295 (1968).
24. W. B. Hardy and R. P. Bennett, *Tetrahedron Lett.* p. 961 (1967).
25. A. F. M. Iqbal, *Helv. Chim. Acta* **55**, 798 (1972); *Angew. Chem., Int. Ed. Engl.* **11**, 634 (1972); *J. Org. Chem.* **37**, 2791 (1972).
26. T. Saegusa and Y. Ito, *in* "Isonitrile Chemistry" (I. Ugi, ed.), Chapter 4. Academic Press, New York, 1971.
27. A. Halleux, *Angew. Chem.* **76**, 889 (1964).
28. J. H. Boyer and W. Beverung, *Chem. Commun.* p. 1377 (1969).
29. J. A. Green and L. A. Singer, *Tetrahedron Lett.* p. 5093 (1969).
30. E. Ciganek, *J. Org. Chem.* **35**, 862 (1970).
31. W. Aumüller, *Angew. Chem.* **75**, 857 (1963).
32. R. Knorr, *Chem. Ber.* **98**, 4038 (1956).

33. A. Krebs and H. Kimling, *Angew. Chem., Int. Ed. Engl.* **10**, 409 (1971).

34. F. Kienzle, *Tetrahedron Lett.* p. 1771 (1972).

35. W. P. Neumann and K. Kühlein, *Tetrahedron Lett.* p. 3423 (1966).

36a. K. Jones and M. F. Lappert, *Organomet. Chem. Rev.* **1**, 67 (1966).

36b. G. Losco, *Gazz. Chim. Ital.* **67**, 553 (1937); M. Passerini and V. Casini, *ibid.* p. 332; M. Passerini, *ibid.* **54**, 633 (1924).

37. R. Corriu, C. Guerin, and J. Masse, *Chem. Commun.* p. 75 (1975).

38. H. Schmidbauer and W. Wolf, *Angew. Chem., Int. Ed. Engl.* **12**, 320 (1973).

39. M. Kumada, K. Tamao, M. Ishikawa, and M. Matsuno, *Chem. Commun.* p. 614 (1968).

40. S. S. Washburne and R. R. Chawla, *J. Organomet. Chem.* **31**C, 20 (1971).

41. A. G. Brook and D. G. Anderson, *Can. J. Chem.* **46**, 2115 (1968).

42. E. Vedejs and M. Mullins, *Tetrahedron Lett.* p. 2017 (1975).

43. R. J. Ouellette, D. L. Marks, D. Miller, and D. Kesatie, *J. Org. Chem.* **34**, 1769 (1969).

44. P. F. Hudrlik and R. Feasley, *Tetrahedron Lett.* p. 1781 (1972).

45. V. Rautenstrauch, *Helv. Chim. Acta* **55**, 3064 (1972).

46. H. Jolibois, A. Doucet, and R. Perrot, *Helv. Chim. Acta* **58**, 1801 (1975).

47. A. G. Brook and P. F. Jones, *Chem. Commun.* p. 1324 (1969).

48. A. G. Brook, D. G. Anderson, and J. M. Duff, *J. Am. Chem. Soc.* **90**, 3876 (1968).

49. A. G. Brook, D. M. MacRae, and W. W. Limburg, *J. Am. Chem. Soc.* **89**, 5493 (1967).

50a. M. T. Reetz, M. Kliment, and M. Plachky, *Angew. Chem., Int. Ed. Engl.* **13**, 814 (1974).

50b. M. T. Reetz and M. Kliment, *Tetrahedron Lett.* p. 2909 (1975).

51. H. Schmidbauer, *Acc. Chem. Res.* **8**, 62 (1975).

52. T. H. Chan, W. Mychajlowskij, and D. N. Harpp, *Tetrahedron Lett.* p. 3511 (1974).

53. G. Stork and E. Colvin, *J. Am. Chem. Soc.* **93**, 2080 (1971).

54. D. R. M. Walton and M. J. Webb, *J. Organomet. Chem.* **37**, 41 (1972).

55. C. Couret, F. Couret, J. Satgé, and J. Escudié, *Helv. Chim. Acta* **58**, 1316 (1975).

56. G. J. D. Peddle and R. W. Waslingham, *J. Am. Chem. Soc.* **91**, 2154 (1969).

57a. T.-L. Ho and G. A. Olah, *Angew. Chem.* **88**, 847 (1976).

57b. T.-L. Ho and G. A. Olah (to be published).

58. D. A. Armitage, M. J. Clark, and A. C. Kinsey, *J. Chem. Soc. C*, p. 3867 (1971).

59. E. W. Abel, J. P. Crow, and J. N. Wingfield, *Chem. Commun.* p. 967 (1969).

60. G. M. Rubottom and M. I. Lopez Nieves, *Tetrahedron Lett.* p. 2423 (1972).

61. W. Adam and J. C. Liu, *J. Am. Chem. Soc.* **94**, 2894 (1972).

62. K. Saigo, A. Morikawa, and T. Mukaiyama, *Chem. Lett.* p. 145 (1975).

63. G. A. Olah and T.-L. Ho, *Synthesis*, p. 609 (1976).

64. U. Weidner and A. Schweig, *Angew. Chem., Int. Ed. Engl.* **11**, 146 (1972).

65. D. J. Peterson and M. D. Robbins, *Tetrahedron Lett.* p. 2135 (1972).

66. H. G. Kuivila, *Synthesis* p. 500 (1970).

67. S. Sakai, Y. Kiyohara, K. Itoh, and Y. Ishii, *J. Org. Chem.* **35**, 2347 (1970).

68. S. Sakai, Y. Kobayashi, and Y. Ishii, *J. Org. Chem.* **36**, 1176 (1971).

69. G. S. Koermer, M. L. Hall, and T. G. Traylor, *J. Am. Chem. Soc.* **94**, 7205 (1972).

70. W. R. Moore, W. R. Moser, and J. E. Laprade, *J. Org. Chem.* **28**, 2200 (1963).

71. C. W. Jefford, nT. Kabengele, J. Kovacs, and U. Burger, *Tetrahedron Lett.* p. 257 (1974).

72. B. Haley, R. N. Haszeldine, B. Hewitson, and A. E. Tipping, *J. Chem. Soc. Perkin Trans. 1* p. 525 (1976).

73. C. L. Osborn, T. C. Shields, B. A. Shoulders, C. G. Cardenas, and P. D. Gardner, *Chem. Ind. (London)* p. 766 (1965).

74. W. R. Moore and H. R. Ward, *J. Org. Chem.* **27**, 4179 (1962); L. Skattebøl and S. Solomon, *Acta Chem. Scand.* **17**, 1683 (1963).

75. C. A. Stein and T. H. Morton, *Tetrahedron Lett.* p. 4933 (1973).

76. D. Merkel and G. Köbrich, *Chem. Ber.* **106**, 2040 (1973).

77. S. Masamune, P. A. Rossy, and G. S. Bates. *J. Am. Chem. Soc.* **95**, 6452 (1973).

78. G. H. Posner and J.-S. Ting, *Synth. Commun.* **4**, 355 (1974).

79. E. J. Corey and G. H. Posner, *J. Am. Chem. Soc.* **89**, 3911 (1967); **90**, 5615 (1968).

80. G. M. Whitesides, W. F. Fischer, Jr., J. San Filippo, Jr., R. W. Bashe, and H. O. House, *J. Am. Chem. Soc.* **91**, 4871 (1969).

81. J. K. Crandall, P. Battuoni, J. T. Wehlacz, and R. Bindra, *J. Am. Chem. Soc.* **97**, 7171 (1975).

82. M. C. Verploegh, L. Donk, H. J. T. Bos, and W. Drenth, *Recl. Trav. Chim., Pays-Bas* **90**, 765 (1971).

83. W. Ried and W. Merkel, *Angew. Chem., Int. Ed. Engl.* **8**, 379 (1969).

84. T.-L. Ho, T. W. Hall, and C. M. Wong, *Synthesis* p. 872 (1974).

85. V. Calo, L. Lopez, G. Pesce, and P. E. Todesco, *Tetrahedron* **29**, 1625 (1973).

86. T.-L. Ho and C. M. Wong, *Synth. Commun.* **3**, 317 (1973).

87. D. B. Denney and F. J. Gross, *J. Org. Chem.* **32**, 3710 (1967).

88. E. C. Ashby, S. H. Yu, and R. G. Beach, *J. Am. Chem. Soc.* **92**, 433 (1970).

89. J. Cantacuzène, D. Ricard, and M. Theze, *Tetrahedron Lett.* p. 1365 (1967).

90. V. Grignard, E. Bellet, and C. Courtot, *Ann. Chim. (Paris)* [9] **4**, 28 (1915).

91. E. H. Bartlett, C. Eaborn, and D. R. M. Walton, *J. Organomet. Chem.* **46**, 267 (1972).

92. J. M. Cox and R. Ghosh, *Tetrahedron Lett.* p. 3351 (1969).

93. C. E. Entemann, Jr. and J. R. Johnson, *J. Am. Chem. Soc.* **55**, 2900 (1933).

94. H. Gilman and J. F. Nelson, *J. Am. Chem. Soc.* **61**, 743 (1939).

95. W. S. Trahanovsky and M. P. Doyle, *Chem. Commun.* p. 1021 (1967).

96. R. G. Pearson, "Hard and Soft Acids and Bases," p. 358. Dowden, Hutchinson, & Ross, Inc., Stroudsburg, Pennsylvania, 1973.

97. R. A. Moss, M. J. Landon, K. M. Luchter, and A. Mamantov, *J. Am. Chem. Soc.* **94**, 4392 (1972).

98. R. F. Hudson, *Angew. Chem., Int. Ed. Engl.* **12**, 36 (1973).

99. C. O. Parker, *J. Am. Chem. Soc.* **78**, 4944 (1956).

100. E. C. Taylor, G. H. Hawks, III, and A. McKillop, *J. Am. Chem. Soc.* **90**, 2421 (1968).

101. L. J. Kricka and A. Ledwith, *J. Org. Chem.* **38**, 2240 (1973).

Addendum

This additional material is intended to bring recent developments to the readers' attention. For easy identification, the following comments are designated by the chapter and specific section to which they pertain.

3.1. Although organosilyl cyanides exist mainly in the normal form, the iso form R_3SiNC is favored at higher temperatures (1). It forms silyl isothiocyanates on heating with sulfur (2).

4.5.1. Regiospecific nucleophilysis is exhibited by carboxylic-phosphinic anhydrides $RCOOPOR_2$. Thus, alcoholysis occurs by attack at P and aminolysis at C (3).

Reaction of CF_3CH_2ONa with $CF_3SO_3CH_2C_3F_7$ takes place at S, and only to a negligible extent at the CH_2 (4).

4.5.2.2. The ambident anion of a 1,3-bisphenylmalondiamidine undergoes exclusive C- or N-methylation, depending on whether the countercation is Li or Na, respectively (5).

4.5.2.3. The attack of the hard, unsolvated *tert*-butyl cation in the gas phase on phenol and anisole is strongly biased in favor of oxygen sites (6).

Treatment of α-diazoketones with $FOCF_3$ leads to mixtures of α-trifluoromethoxy-α-fluoroketones and α,α-difluoroketones (7). The reaction involves successive attacks by $F^\oplus$ and $CF_3O^\ominus/F^\ominus$. The $F^\ominus$ ion comes from $CF_3O^\ominus$.

4.5.2.4. The ambident lithium (trimethylsilylethynyl)thiolate reacts with Me_3SiCl to give $(Me_3Si)_2C=C=S$ and with the softer Me_3SiBr to afford $Me_3SiC\equiv CSSiMe_3$ (8). The acetylenic sulfide is thermally isomerizable to the thioketene, indicating that linking of S simultaneously to hard Si and *sp*-hybridized carbon atoms is much less favorable.

182

Polylithiopropargylides attack softer acids with C-3 and very hard acids with acetylide carbons (9).

5.1. That alkyl phenyl selenoxides decompose at or below room temperatures (10) is in keeping with the very soft nature of Se, which prefers being at low oxidation states.

6.3. The glutacondialdehyde anion reacts with isothiocyanates to give 3-formylpyridinethiones (11). The initial C—C bond formation involves interaction between the relatively soft centers, and cyclization is a hard–hard interaction of N and C=O. (Following ref. 56.) It has been shown that trivalent N, as in $H_2NOSO_3^-$, is a softer electrophile than carbon (12).

Nitrosoarenes form N-hydroxysulfonamides with arenesulfinic acids (13) through a soft–soft interaction.

8.3.2. 4-Hydroxyarsabenzene is acylated at oxygen but alkylated at the softer arsenic site (reactivity EtI > EtBr > EtOTs) (14).

9.1. Degradation of α-silylselenides $RCH(SiMe_3)SePh$ by H_2O_2 involves Ⓢ: Ⓢ-//-Ⓗ: Ⓗ interaction to generate trimethylsilanol and hydroxyselenonium ylides. The latter species react with a second molecule of H_2O_2, furnishing the unstable α-hydroxy selenoxides which decompose to aldehydes and PhSeOH (15). All steps conform to Saville's rule.

The reactivity of $NCTe^\ominus$ > $NCSe^\ominus$ > $NCS^\ominus$ toward benzyl bromide (16) follows their relative softness.

Phosphines efficiently abstract selenium from both inorganic (17) and organic isoselenocyanates (18).

9.2. The reaction of dialkoxysulfides with alkyl iodides to give alkanesulfinic esters (19) is analogous to the Arbuzov reaction.

2-Arylaminopyrylium salts are readily converted into 2-pyridones, whereas the corresponding thiopyrylium salts give thiopyranimines only (20). This is another example showing the reluctance of divalent sulfur to form a double bond with carbon.

Certain benzyl sulfides are converted to the benzyl methyl ethers by reaction with thallium(III) nitrate in methanol (21). The soft donor characteristic of thiocarbonyl and selenocarbonyl groups has been exploited in the replacement of hydroxy functions with halogen atoms (22). Thus, xanthates and thiobenzoates are activated by alkyl halides.

9.3. Grignard reactions on thiones are exclusively thiophilic even in the presence of carbonyl groups (23–25).

Cleavage of the thiolsulfinate t-BuS-S(O)NMe$_2$ can be effected by the soft acid $HgCl_2$ (26).

The sulfenic acid generated by thermolysis of a penicillin sulfoxide has been trapped as the O-trimethylsilyl derivative (27), whereas alkanesulfenic acids

(28) and *tert*-butanethiosulfoxylic acid (29) add to $\alpha\beta$-unsaturated esters using their S-termini.

The 2-nitro-4-trifluoromethylbenzenesulfenate ion is S-methylated with MeI, but it gives predominantly the O-methyl product with $MeOSO_2F$ (30).

The formation of $Me_2S=\overset{\oplus}{N}=SMe_2$ $X^{\ominus}$ from the reaction of DMSO with cyanogen halides (31) must be initiated by attack of the sulfoxide oxygen on the nitrile carbon.

10.1.1. The cleavage of methyl ethers by a combination of $Et_2O \cdot BF_3$ and thiols has been reported (32).

11.1. para-Protonation of toluene appears to be frontier controlled according to charge calculations (33). Approach of the proton to the ortho positions might suffer repulsion from the soft methyl group.

11.4. Alkyl iodides are more efficiently prepared by displacement of tosylates with magnesium iodide (34) than sodium iodide. It is HSAB relevant. The ten rules formulated by Gilman (35) concerning reactivities of organometallics can be qualitatively explained by considering the degree of hard–soft dissymmetry of the metal and carbon atoms, and the higher reactivities of unsymmetrical RMR' in comparison with the symmetrical R_2M might be due to symbiosis.

REFERENCES

1. T. A. Bither, W. H. Knoth, R. V. Lindsey, Jr., and W. H. Sharkey, *J. Am. Chem. Soc.* **80**, 4151 (1958).
2. J. J. McBride, Jr., and H. C. Beachell, *J. Am. Chem. Soc.* **74**, 5247 (1952).
3. A. G. Jackson, G. W. Kenner, G. A. Moore, R. Ramage, and W. D. Thorpe, *Tetrahedron Lett.* p. 3627 (1976).
4. P. Johncock, *J. Fluorine Chem.* **4**, 25 (1974).
5. S. Brenner and H. G. Viehe, *Tetrahedron Lett.* p. 1617 (1976).
6. M. Attina, F. Cacace, G. Ciranni, and P. Giacomello, *Chem. Commun.* p. 466 (1976).
7. C. Wakselman and J. Leroy, *Chem. Commun.* p. 611 (1976).
8. S. J. Harris and D. R. M. Walton, *Chem. Commun.* p. 1008 (1976).
9. W. Priester and R. West, *J. Am. Chem. Soc.* **98**, 8421 (1976).
10. K. B. Sharpless, K. M. Gordon, R. F. Lauer, D. W. Patrick, S. P. Singer, and M. W. Young, *Chem. Scripta* **8A**, 9 (1975).
11. J. Becher and E. G. Frandsen, *Tetrahedron Lett.* p. 3347 (1976).
12. J. H. Krueger, B. A. Sudbury, and P. F. Blanchet, *J. Am. Chem. Soc.* **96**, 5733 (1974).
13. A. Darchen and C. Moinet, *Chem. Commun.* p. 820 (1976).
14. G. Märkl and J. B. Rampal, *Tetrahedron Lett.* p. 4143 (1976).
15. K. Sachdev and H. S. Sachdev, *Tetrahedron Lett.* p. 4223 (1976).
16. T. Austad, S. Esperas, and J. Songstad, *Acta Chem. Scand.* **27**, 3594 (1973).
17. P. Nicpon and D. W. Meek, *Inorg. Chem.* **5**, 1297 (1966).
18. L. J. Stangeland, T. Austad, and J. Songstad, *Acta Chem. Scand.* **27**, 3919 (1973).
19. E. Wenschuh, R. Fahsl, and R. Höhne, *Synthesis* p. 829 (1976).

20. A. S. Afridi, A. R. Katritzky, and C. A. Ramsden, *Chem. Commun.* p. 899 (1976).
21. Y. Nagao, K. Kaneko, M. Ochiai, and E. Fujita, *Chem. Commun.* p. 202 (1976).
22. D. H. R. Barton, R. V. Stick, and R. Subramanian, *J. Chem. Soc., Perkin Trans. 1*, p. 2112 (1976).
23. P. Metzner, J. Vialle, and A. Vibet, *Tetrahedron Lett.* p. 4295 (1976).
24. J.-L. Burgot, J. Masson, P. Metzner, and J. Vialle, *Tetrahedron Lett.* p. 4297 (1976).
25. J.-L. Burgot, J. Masson, and J. Vialle, *Tetrahedron Lett.* p. 4775 (1976).
26. M. Mikolajczyk and J. Drabowicz, *Chem. Commun.* p. 775 (1974).
27. T. S. Chou, *Tetrahedron Lett.* p. 725 (1974).
28. E. Block, *J. Am. Chem. Soc.* **94**, 642 (1972).
29. E. Block, *J. Am. Chem. Soc.* **94**, 644 (1972).
30. D. R. Hogg and A. Robertson, *Tetrahedron Lett.* p. 3783 (1974).
31. P. Y. Blanc, *Experientia* **21**, 308 (1965).
32. M. Node, H. Hori, and E. Fujita, *J. Chem. Soc., Perkin Trans. 1*, p. 2237 (1976).
33. O. Chalvet, C. Decoret, and J. Royer, *Tetrahedron* **32**, 2927 (1976).
34. J. Gore, P. Place, and M. L. Roumestant, *Chem. Commun.* p. 821 (1973).
35. H. Gilman, "Organic Chemistry, an Advanced Treatise," Vol. 1, pp. 521–524. Wiley, New York, 1943.

Author Index

Numbers in parentheses are reference numbers and indicate that an author's work is referred to although his name is not cited in the text. Numbers in italics show the page on which the complete reference is listed.

Subject Index

A

Acetals, 22, 126
Acetoacetates, enolization, 15
Acetolysis, 168, 177
Acetylcholinesterase, 118
Acids and bases
  Brønsted-Lowry, 1
  hard and soft, 2, 5ff.
  Lewis, 1
Acylferrocenes, protonation, 163
Acyloins, 90
Acyloxyhalides, 13
Acyloxysilanes, organometallic reactions, 168
Aldol condensation, 15, 89
Alkoxydiazenium ions, 36
α-Alkoxy-β-oxosulfone anion, 48
Alkylamines, from trialkylboranes, 153
Alkylation
  pressure effect, 41
  two-phase, 29
α-(Alkylthio) ketones, dehydrogenation, 132
(Alkylthio) sulfonium salts, 61
Allopolarization, 40
Allyl carbanion, 49
Allylic alcohols, from epoxides, 159
Allylsilanes, 172
Ambident electrophiles, 35
Ambident nucleophiles, 39ff.
Ambident reactivity, 35, 80, 120ff., 127ff., 139, 143, 177
Amide, protonation, 40
Amide oximes, 41
Amination, 46
Amino acid, synthesis, 157
Aminodefluorination, 74
Aminometallation, 167
Aminophosphines, 17
Anhydro sugar, 126

Anion radical, K^+-encapsulated, 162
Anthrone, alkylation, 46
Arbuzov reaction, 105ff., 117
Arylboron, 157
Aryldimethylsilane, oxidation, 168
Arylhydrazines, from diazonium salts, 144
"Ate" complexes, 159
Azetidinones, 38
Azides, reduction, 20
Azo salts, 36

B

Back-donation, 19
Baeyer–Villiger oxidation, 87ff
Baker–Nathan effect, 71
Basicity, 27
Benzophenone dianion, 46
Benzyne, 76, 104, 144, 166
Bidentate, 57
Biphilicity, 20, 103, 110, 111, 113, 131, 132, 134, 137, 140, 144
Birch reduction, 80
Bond contraction, 21
π-Bonding theory, 9
Borane–alkene complexes, 155
Boranides, 158
Boron trihalides, 151
2-Bromoethyl esters, alkylation, 178
Bunte salts, 64, 115, 137

C

Carbamates, 40, 167
Carbanion(s)
  from haloforms, 64
  vinyl, 65
Carbanion–carbene complex, 80
Carbene, 9, 20, 60, 105, 112, 133, 143, 154, 164

205

HARD AND SOFT ACIDS AND BASES PRINCIPLE IN ORGANIC CHEMISTRY

FROM THE PREFACE:

The founding of a coherent concept always signifies major advances in scientific endeavors. In this spirit the *Hard and Soft Acids and Bases (HSAB) Principle* was promulgated by Professor R. G. Pearson more than a dozen years ago. In the meantime a gargantuan body of chemical phenomena related directly or implicitly to HSAB has been accumulated and clarified, most of which vindicates Pearson's viewpoint. As it stands, there is no gainsaying the value of the Principle as a convenient yet powerful rule of thumb for assessing and predicting chemical events without resorting to lengthy and cumbersome calculations. As the HSAB principle offers great assistance and stimulation for both pedagogy and research, it should be introduced to students of chemistry at an early stage.

Starting from explaining inorganic coordination chemistry HSAB has subsequently been applied to almost every chemical ramification with considerable success. Although pioneering applications to the organic area by Hudson, Saville, and Pearson himself have irrefutably demonstrated the relevance of HSAB to organic chemistry, subject matter hitherto touched upon is likened to the exposed tip of an iceberg. Moreover, pertinent findings are bound to surface as long as organic research does not cease.

The present volume represents an attempt to examine many organic facets within the HSAB context.